SUBACUTE CARE

Redefining Healthcare

LAURA HYATT

 HFMA® HEALTHCARE FINANCIAL MANAGEMENT ASSOCIATION

A Healthcare 2000 Publication

 IRWIN
Professional Publishing®
Burr Ridge, Illinois
New York, New York

A **2000** *PUBLICATION*

ISBN 1-55738-630-7

Printed in the United States of America

BSPC

1 2 3 4 5 6 7 8 9 0

CB/BJS

To Dave
*"That's the way it was
movin' west."*

Table of Contents

Foreword

By Art Linkletter

Healthcare used to be so simple. In the "old" days, you went to your family doctor. If you needed to see a specialist, your doctor made a referral. If you got very sick, you went to the hospital. When you got better, you went home. If you couldn't be cared for at home, you went to a nursing home. Your insurance covered everything.

With advances in medical technology today, healthcare is moving in new directions. Several new trends offer positive solutions to today's more complex health service problems. I recommend that you read this book, *Subacute Care*, which is dedicated to a positive healthcare approach. Subacute care is a level of care between the hospital and the nursing home. It is usually provided in a freestanding nursing home or in a nursing home within a hospital. It involves many kinds of healthcare professionals and high technology. It might be best likened to an intensive care unit within a nursing home.

To begin to understand how subacute care fits into the healthcare system and how you might benefit from this trend, you should consider the driving forces that led to the subacute trend. Incredible advances in medical technology are saving lives every day that just a few years ago would not have been possible. Laproscopic surgery, for example, allows many people to return home after a one- to two-day stay instead of surgeries that once required weeks of hospitalization. These advances and increased specialization in medicine have allowed many persons to live with complex medical conditions. Although stabilized in an acute care part of the hospital within a matter of days, these patients may have more complex medical and rehabilitation needs that require further attention. For many of these patients, subacute care is the answer.

The aging of America has allowed a growing number of persons to celebrate more birthdays than ever before. The 85 and older age segment is the fastest growing part of the population. However, with advanced age often come more complicated medical problems. Often these problems may arise because of preexisting conditions, such as diabetes, heart disease, and arthritis. Subacute care programs are now helping many Americans who have such multiple medical problems.

Changes in how healthcare is paid for and administered have also been a driving force in the development of the subacute level of care. Today many of us are part of the alphabet soup of managed care and are members of health maintenance organizations (HMOs) or preferred provider organizations (PPOs) that may direct us to physicians, specialists, hospitals, subacute care programs, skilled nursing facilities, outpatient clinics, and other healthcare providers that are part of their network. Even those who use traditional health insurance are more cost-conscious than ever. Skyrocketing healthcare costs have put a new focus on finding the most appropriate setting of care that offers both cost savings and the best chance for rehabilitation and recovery. That focus has led to an increased use and recognition of subacute care, which typically offers significant cost savings while providing focused care for specific medical problems.

What does this mean for you? It means that there is an alternative that will provide for your medical needs and assist in extending healthcare benefits. These programs offer advanced technologies but, perhaps more importantly, they also offer highly specialized medical professionals with skills and experience. Usually these programs offer a more homelike environment to foster recuperation. Visiting hours are more liberal compared with the traditional hospital, and families and loved ones are encouraged to become involved. Often these programs offer common dining areas and activities to encourage a return to the routine of daily life.

In short, such subacute care programs are offering healthcare based on patients' needs. The result is more appropriate quality care with significant cost savings. Subacute care programs are typically 20 to 40% less expensive than traditional hospital care. Much of the savings comes from the fact that patients are no longer paying for the hospital overhead involved with services that are not relevant to their needs, such as emergency rooms or operating theaters. Americans need to know that billions can and should be saved through greater use of subacute care programs.

Subacute programs include care for a variety of diagnoses including rehabilitation, oncology, infectious diseases, orthopedic injuries, cardiac disease, respiratory problems, and strokes. I mentioned that subacute care is helping many older Americans who suffer from multiple medical conditions, but I should add that there are a growing number of pediatric subacute care programs offering new and better choices for thousands of children with complex medical needs.

By the end of the century, I suspect that most American families will be helped at some point by a subacute care program. As we look for fresh approaches to solving today's healthcare dilemma, subacute care has already emerged as one important part of the solution.

Art Linkletter
Entertainer/
Vice-Chairman, Friends of the Center on Aging
University of California at Los Angeles/
Member of the Board of Directors of Capp Care

Foreword

By Captain Walter "Wally" Schirra

Everyone is talking about healthcare—the method in which it is delivered and how expensive it is. I believe most people are concerned with receiving quality services that are available and affordable. Yet, our healthcare needs seem to keep growing while costs keep escalating.

Just when things seemed as though they were on a collision course, Americans summoned up the ingenuity and resourcefulness that they are famous for and created "subacute care." This is a level of care that can be provided in a hospital or in a nursing home, that meets patient's needs and pocketbooks without sacrificing quality.

Upon examination it provides a solution to many of the problems that are facing America as well as many other countries around the world, and when I say around the world—I say it with some experience.

Insurance companies and physicians have also discovered subacute care and its benefits. Many are admitting patients directly to subacute care programs from home if they do not require surgery or diagnostic work-ups. These programs are less expensive, simply because they have fewer costs associated with running them. They are more like mini-hospitals with doctors and nurses and a full range of therapists.

These are professionals who work as a team, often with special training and experience, to address a more focused approach to care. You can find high-technology monitoring devices that includes intravenous, cardiac, and pulmonary equipment to name a few.

Many subacute care programs are directed at specific illnesses such as cancer, strokes, heart disease, neurologic, orthopedic, Alzheimer's disease, infections, and postacute recovery.

These facilities are often more comfortable than are traditional hospitals. Patients are encouraged to have some personal items in their rooms; the decor is homelike; and families are actually desired guests and often experience more flexible visiting hours.

People often ask me if I had always wanted to become an astronaut. The truth is I had not thought much about space flight at all. However, after hearing the briefings at NASA, I experienced the same instinct that inspired me to become a test pilot. This was to improve flying and to expand the frontiers so that the whole nation would benefit. The same applies to the medical professionals who are pioneering new and better ways to care for people—we will all reap the rewards from their efforts.

We all need to learn more about subacute care. Laura Hyatt has written a book that enlightens us and gives us hope for the future. I hope you read *Subacute Care.* If you are a healthcare professional, the book may encourage you to develop a subacute care program; if you are a business person it might present you with new opportunities; if you are an insurer it will offer you an option that can assist your clients and extend their benefits; if you are a professor of medicine, nursing, therapies, health or business administration, it should be required reading for your students; and if you are a consumer it will help you to become better informed about your healthcare choices.

Captain Walter "Wally" Schirra
Astronaut/
Member of the Board of Directors of Sharp HealthCare Foundation

Acknowledgments

I would like to express my appreciation for the interest and contributions of the persons listed here. The willingness to share their expertise is evidence of the commitment that they bring to this industry. In addition I want to acknowledge Marie Scimeca for her computer skills and diligence to the task. Many thanks are owed to my friends, colleagues, professors, and associates who have offered information, suggestions, patience and most of all encouragement.

It is my hope that this work will assist in contributing to the advancement of the healthcare field, as well as provide a resource for the challenges of the future.

CONTRIBUTORS

Armand E. Balsano
Partner
Gill/Balsano Consulting

Mark G. Banta, CFA, JD
Vice-President
Analyst-Health Services
Salomon Brothers, Inc.

Maureen L. Barry, RN, BSN, CCM
Medical Management
Managed Indemnity/Preferred Provider Organizations Programs
MetLife

Kevin Cornish, RRT, BS
Assistant Administrator Outpatient Services
Vencor

Robert Craig
Chief Operating Officer
Covenant Care, Inc.

Dennis S. Diaz, Esq.
Partner
Musick, Peeler & Garrett
Professor of Law
University of Southern California

Harriet S. Gill
Partner
Gill/Balsano Consulting

Richard S. Goka, MD
Chair, Subacute Care Task Force
Academy of Physical Medicine and Rehabilitation

Clair Jones, MS
Director for Rehabilitation Continuum Services
Sharp HealthCare

Michael W. Kesti
Director of Marketing and Business Development
Seniors Management

Pamella Leiter, MSA OTR/L
President
Formations in Healthcare, Inc.
Assistant Professor
DePaul University

Christine M. MacDonell
National Director Medical Rehabilitation Division
Commission on Accreditation of Rehabilitation Facilities (CARF)

Jerald Moore
Executive Vice-President
Beverly Enterprises

Samantha Morgan, RN, BSN, CCRN, CETN, CCM
Morgan Morgan & Associates

Mary Tellis-Nayak, RN, MSN, MPH
Director Long Term Care Accreditation Services
Joint Commission on Accreditation of Healthcare Organizations (JCAHO)

John P. Perticone, MBA, CPCU, CEBS
Corporate Director of Marketing & Case Management
The Multicare Companies, Inc.

Introduction

*"You must make yourself contented by
the consciousness of success."*

— Dr. Watson

The ubiquitous state of healthcare institutions is changing. Payors are encouraging a unification of resources in favor of forming integrated delivery systems. Providers are redefining healthcare services. Subacute care is a significant component in this new equation.

As we enter the next medical millennium, healthcare will require a shift in paradigms. We will have to change our standard patterns of thought and behavior with regards to healthcare. Less energy is required to change a standard thinking pattern. It is usually accomplished through education, motivation, exposure, and experience. However, a change in opinion may have no bearing on our actions. Changing behavior is much more complex and requires a greater amount of effort and energy.

For example, a healthcare facility can be viewed as an entity in itself. This attitude usually ignores personnel turnover largely because it is the position, as opposed to the staff member, that is the focus. Another common way of looking at a healthcare facility treats the position as one stage of an occupational career, so that turnover is increased and the organization is seen as a brief stop on the way to somewhere else. The process by which care is rendered and the quality of those services has everything to do with the people delivering the services, and less to do with the bricks and mortar of the building. Subacute care is an approach to healthcare that promotes an interdisciplinary team effort, nurtures a sense of value, and encourages the retention and continuity of its staff members.

The subacute care model creates a more user-friendly, more personal, less autocratic care setting for the patient as well. Often a consumer admitted to a hospital feels powerless, vulnerable, and insignificant. Many people who are hospitalized never see the same caregiver twice. These patients are probed, medicated, transported, and examined by personnel with whom they have little opportunity to develop a relationship. They become an illness or an injury, that is, the fracture in room 405 or the kidney failure in bed "B." Subacute programs are putting the "care" back in healthcare. By treating patients as complete people and including them and their family or loved ones as part of the care team, a sense of empowerment is instilled which in turn fosters recovery.

The evolution of the subacute care program is an innovative response to a need for quality healthcare that controls costs. By establishing these care capabilities in a setting which is not encumbered by the high overhead costs of the hospital, subacute care programs are able to offer comparable, often more focused services at a lower price.

I expect, in addition to subacute care, that many more innovative ways to provide medical services are just around the corner. They will come from consumers and academicians, as well as professionals, as we strive to improve the present systems and redefine healthcare.

PART I

◆

SUBACUTE CARE
PERSPECTIVES

CHAPTER 1

Redefining Healthcare

"Truth has no special time of its own. Its hour is now . . . always."
— Dr. Schweitzer

People are living longer, more productive lives. Improvements in medical technology are increasing the survival rate of victims of catastrophic injury or illness. This is the age of information that is readily available to the consumer. The world is changing and so is medical care. It is time to reevaluate and redefine the way we deliver healthcare services. It is important to develop methodologies for providing quality care that are both efficient and effective as part of an integrated process. Subacute care will play an important role and be a valued contributor in redefining healthcare.

Subacute care programs offer value enhancement to an integrated delivery system. A subacute care capability enables a system to provide another service option. Additionally, subacute care is a cost-effective alternative that permits the system to operate more efficiently and to compete for managed care contracts.

Integrated delivery systems can coordinate a continuum of healthcare services for the payor and the patient. Satisfying the need for services and geographic coverage, integrated delivery systems can respond to market changes because they include several different types of providers:

- Acute care
- Subacute care
- Long-term care
- Home health care
- Ancillary services
- Outpatient care

- Assisted living
- Day treatment

Simultaneously, the integrated delivery system is in a position to lower costs by reducing or eliminating duplicate services and achieving economies of scale among all of its entities. In addition, providers within the system can coordinate services together better than they can separately. Subacute care is an important addition to the integrated delivery system's cost-cutting efforts because it meets the needs of the system's efficient care provisions.

As managed care increases and health maintenance organizations (HMOs) experience rapid growth, there will be an even greater demand for subacute services. Recently, HMO membership has increased substantially because of the addition of Medicare HMOs or senior plans. There are two basic types of Medicare HMOs: *cost contract HMOs* that are paid the actual costs of providing care; and *risk contract HMOs* that are paid a capitated rate per Medicare enrollee in return for providing Medicare-covered services. HMOs are spending millions of dollars on television and radio commercials, print advertisements, direct mail, and telemarketing to seniors. The popularity of these plans is on the rise. For instance in California between 1989 and 1991, Medicare HMO enrollment grew at the rate of 20% per year. This represents about a million people in just one state. Similarly, there has been increased recognition and utilization of subacute care programs by the Medicare HMOs. Managed care organizations such as HMOs do not require a three-day hospital stay before a patient can be transferred to a skilled nursing facility. Subacute care programs are often housed in skilled nursing facilities. As the number of Medicare HMO enrollees increases, subacute care programs will be the recipients of considerable benefits.

Presently, many payors—including HMOs—have policies in effect for non-Medicare enrollees that encourage bypassing acute care entirely, in favor of direct admission from home to a subacute care program.

DEFINITION OF SUBACUTE CARE

As of this writing, there are a multitude of definitions available for subacute care. Rapidly increasing in acceptance is the one that was developed by the Joint Commission on the Accreditation of Healthcare Organizations (JCAHO):

> Subacute care is comprehensive inpatient care designed for someone who has had an acute illness, injury, or exacerbation of a disease process. It is goal-oriented treatment rendered immediately after, or instead of, acute hospitalization to treat one or more specific active complex medical conditions or to administer one or more technically complex treatments, in the

context of a person's underlying long-term conditions and overall situation.

Generally, the individual's condition is such that the care does not depend heavily on high-technology monitoring or complex diagnostic procedures. It requires the coordinated services of an interdisciplinary team including physicians, nurses, and other relevant professional disciplines, who are trained and knowledgeable to assess and manage these specific conditions and perform the necessary procedures. Subacute care is given as part of a specifically defined program, regardless of the site. Subacute care is generally more intensive than traditional nursing facility care and less than acute care. It requires frequent (daily to weekly) recurrent patient assessment and review of the clinical course and treatment plan for a limited (several days to several months) time period, until a condition is stabilized or a predetermined treatment course is completed.

One of the main reasons that acceptance of a definition has been delayed revolves around recognition by the government. Various governmental entities are currently studying subacute care in order to better define and subsequently reimburse for subacute care services.

To understand this, it's necessary to review the federal government's definitions of two other types of facilities.

The following are definitions according to Title XVIII:

Skilled Nursing Facility

"The term skilled nursing facility means an institution (or distinct part of an institution) which—

(1) is primarily engaged in providing to residents—

(A) skilled nursing care and related services for residents who require medical and/or nursing care, or

(B) rehabilitation services for the rehabilitation of injured, disabled, or sick persons, and is not primarily for care and treatment of mental diseases"

Hospital

"The term hospital (except for purposes of sections 1814(d), 1814(f), and 1835(B), subsection(a)(2) of this section, paragraph (7) of this subsection, and subsection (i) of this section) means an institution which—

(1) is primarily engaged in providing, by or under the supervision of physicians, to inpatients (A) diagnostic and therapeutic

services for medical diagnosis, treatment, and care of injured, disabled, or sick persons, or (B) rehabilitation services for the rehabilitation of injured, disabled, or sick persons . . ."

Let's examine these definitions more closely to distinguish the skilled nursing facility (SNF) from the hospital:

1. The terminology in the hospital definition of "by or under the supervision of physicians" is also true for SNFs although not stated in this section.

2. The words that describe the users of these two entities differ: "residents" for SNFs and "inpatients" for hospitals. This may have been true at one time, but presently skilled nursing facilities care for both "residents" who may spend the remainder of their lives there and "inpatients" who for short periods of time require treatment and technology unavailable to them elsewhere.

3. The most obvious difference is the reference that hospitals provide services for medical diagnosis. This, as a rule, is a significant distinction between hospitals and SNFs. However, it is common for skilled nursing facilities to provide both laboratory and x-ray services that are used to both track and diagnose many medical situations.

The first two differences are not presently accurate, so let's remove that specific verbiage and look at the two definitions again.

EXHIBIT 1.1
HOSPITAL VERSUS SKILLED NURSING FACILITY

	HOSPITAL		*SKILLED NURSING FACILITY*
(1)	is primarily engaged in providing . . . , to . . .	(1)	is primarily engaged in providing to . . .
(A)	diagnostic services and therapeutic services for medical diagnosis, treatment, and care of injured, disabled, or sick persons, or	(A)	skilled nursing care and related services for . . . who require medical and/or nursing care, or
(B)	rehabilitation services for the rehabilitation of injured, disabled, or sick persons	(B)	rehabilitation services for the rehabilitation of the injured, disabled, or sick persons

---

The author recognizes that there are numerous other state and federal regulations to which these entities have to adhere. However, the basic federal definition is very similar for both skilled nursing facilities and hospitals. To take this a step further, it stands to reason that if there are few notable differences in the definitions of SNFs and hospitals, no wonder the government is having a difficult time defining subacute care. Additionally, it appears that indemnity and managed care organizations are having an easier experience recognizing, contracting, and reimbursing subacute care programs for services to their enrollees.

THE BEGINNING

The history of subacute care has not been well recorded. It is even less accurately reported. Most articles state that the subacute care field began in the mid-1980s. This is incorrect or maybe it's a "regional truth." Subacute care programs actually existed under that name as early as 1976. The misinformation was caused by a few large publicly traded companies who wanted to capture the cash, the clients, and the credit for creating subacute care. In the process, however, they began to believe their own propaganda.

In 1976 a nurse from a Massachusetts skilled nursing facility with a strong rehabilitation component was sent to Bellevue Hospital in New York to perform an assessment of three patients. Two of the patients were elderly and had suffered a stroke; the third was an 18-year-old with a catastrophic head injury. These patients had medical and rehabilitative needs and were located in the "disposition unit." This unit was specifically for patients who presented placement problems.

At this time New York City, like all large urban areas, was experiencing a growth in violent crime. Victims of violence often required longer periods of recovery. Concurrently, improvements in technologies and medical care had more than doubled the rate of survival. Also, it was not uncommon for a city such as New York to send patients to facilities in other states in order to recover. This was especially true if the services were unavailable or cost-prohibitive.

After completing the assessments, the nurse returned to Massachusetts to discuss the findings with the interdisciplinary team. The team decided to accept the two elderly patients. However, when they called Bellevue to notify the hospital of their decision, they were told that they must accept all three patients or none.

Fortunately, the facility had been providing a subacute care program with a highly skilled interdisciplinary team of professionals. The administrator telephoned Rancho Los Amigos, a rehabilitation facility in California that had published a scale defining the different functional levels of a brain-injured patient. Today this measurement is commonly referred to as the "Rancho Scale."

One of the authors of this scale spoke with the facility and reaffirmed the level of the patient. Encouraged by this information, the Massachusetts facility accepted all three patients. The head-injured patient eventually progressed and was able to reenter the community.

A doctor with New York City Health Services worked with the Massachusetts facility to develop a special reimbursement rate. Then a plan was implemented to care for patients who had survived a catastrophic event in numbers large enough to comprise a group.

This event is significant to the beginnings of subacute care because it meets all of the following 10 criteria:

1. It was called subacute care
2. Beds utilized were licensed for other than acute care
3. An organized program was in place
4. There was a medical director with special training dedicated to the program
5. An interdisciplinary team was utilized
6. Specific admission criteria were based on medically complex and/or rehabilitative needs
7. Goals and objectives were outcome focused
8. A distinct part of the facility was used
9. The reimbursement rate was specific to the program, not the facility.
10. The goal was to return the client to the community

Initially subacute care was spawned by the rehabilitation industry and increased effectiveness of medical technologies. The ability to perform heroics and save the lives of patients who had suffered catastrophic illness or injury created an entirely new patient population. Often these patients were admitted with a unique set of complications requiring extensive medical and rehabilitative services. The length of their recovery period often extended beyond weeks to months. The aforementioned patients could not tolerate the prescribed three hours of therapy necessary in order to occupy an acute rehabilitation bed. Families were unable to care adequately for their loved ones or provide the necessary services that they required. These patients could not remain in the acute hospital bed and the majority of skilled nursing homes did not have the equipment or trained staff in place to treat them. Insurance carriers were desperately seeking relief as a number of these patients would require lifetime care.

Simultaneously, another phenomenon was occurring: The elderly population was growing in numbers at a rapid rate. In the meantime, the healthcare industry was reeling from the impact of the Omnibus Budget Reconciliation Act

(OBRA) and the federal government's payment system. Patients were being discharged after a brief hospital stay, and acute care facilities were scrambling to find an appropriate setting for disposition.

The most obvious sites turned out to be both hospital-based and freestanding skilled nursing facilities (SNFs). After all, this setting was often used to extend the recovery period for patients who could not go directly home, right? Wrong! At first this plan of action was a disappointment. The SNF was not prepared to accommodate the medically complex patient. There hadn't been an opportunity to hire nurses and therapists with the skills and experience to provide for this higher-acuity patient population. They were ill-equipped technologically and had few to no structured programs in place. This resulted in patients regressing or incurring complications such as pneumonia and requiring rehospitalization. As expected, this process ultimately cost more money.

Although unsuccessful at first, many SNF providers had sampled a higher reimbursement rate. This was because of the patients' increased service needs. As the saying goes, "How are you going to keep them down on the farm after they've seen Paris?" The SNF providers realized that they ultimately would be responsible for these patients and that this situation presented a revenue opportunity.

FACTORS THAT LED TO THE CREATION OF SUBACUTE CARE

- Decreased lengths of stay in the acute hospital
- Payors demanding a different venue at a lower reimbursement rate
- The increase in survival from catastrophic illness or injury
- A rapidly growing elderly population
- Changes in Medicare
- A high touch-high tech capability at the skilled nursing facility
- The proliferation of managed care organizations
- More educated consumers
- Economics

THE PROFESSIONALS

I have often maintained that any business is only as good as its employees. The interdisciplinary team is the most important component of any subacute care program, and any corporate executive who doesn't believe that is guaranteed to experience failure. To implement a subacute care program, there must be a complete understanding of the team, its purpose, function, and the various roles involved.

The subacute care interdisciplinary team's responsibility is to provide physical treatment to assist the patient to achieve an optimal level of recovery and independence. The team is also responsible to the facility administration and the parent organization to follow policies and procedures, promote the ethical values and mission, and actively participate in achieving the economic goals of the organization as a whole.

Subacute care personnel require ongoing training and team building to enable them to transcend their learned adherence to a discipline or department. Communication is crucial and each team member should be encouraged to share information with all others.

Every subacute care program should evaluate necessary staff specific to the program and the patient's needs. However, there are some positions that are integral to most subacute care programs. A sample organizational chart appears in Exhibit 1.2. Some of these team members will be discussed in detail.

Medical Director

The medical director should be a physician—preferably with experience, special training, and board certification in the area of focus of the subacute care program. Physician visits should be as frequent as indicated by the patient's medical condition. The physician is ultimately responsible for the plan of care. The medical director is essential as well to utilization of the subacute care program. A sample job description for a subacute care medical director is presented in Exhibit 1.3.

Administrator

The administrator must have obtained appropriate licensure and credentials. The administrator should have knowledge of subacute care in addition to knowledge related to the facility licensure. There must be a thorough under-standing of the different reimbursement mechanisms. Medicare, Medicaid, indemnity, managed care, capitation, and worker's compensation all have different requirements and criteria.

The administrator must be flexible and adaptable. It is most necessary to provide leadership and set an example. An administrator should always do what is most appropriate for the patient regardless of the economic pressures that may

EXHIBIT 1.2
SAMPLE OF AN ORGANIZATION CHART FOR A SUBACUTE CARE PROGRAM

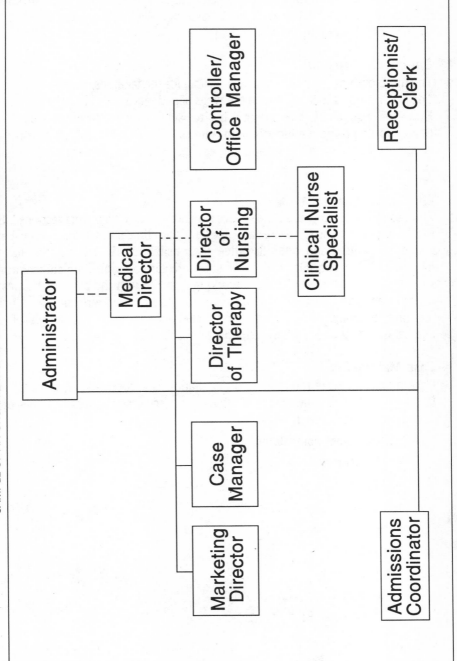

occur. The administrator must have a clear understanding and keep current on all local, state, and federal regulations that impact the facility's licensure and the subacute care program.

Director of Nursing

The director of nursing (DON) is responsible for the facility's nursing personnel. The staff required will depend on the types of patients admitted to the programs. However, a subacute care program has registered nurses 24 hours a day. The DON should promote education and training. The DON is also responsible for keeping current on all laws and regulations that effect the facility's nursing service.

Clinical Nurse Specialist

This registered nurse, under the supervision of the DON, has direct responsibility for patient care in the subacute care program. The clinical nurse specialist usually has both clinical and administrative responsibilities. In addition to formal education in the specialty area of the subacute care program, the clinical nurse specialist should have experience working directly with the particular types of patients admitted to the subacute care program. An understanding of team building, outcomes, and organizational skills is imperative. A sample of the clinical nurse specialist job description is presented in Exhibit 1.4.

Case Manager

Case management is the process of facilitation and coordination of care. The facility-based case manager for the subacute care program is usually a registered nurse with experience in

- Organizational systems
- Management
- Clinical skills
- Administrative responsibilities
- Payor knowledge
- Assessment
- Outcomes
- Report writing
- Financial abilities
- Negotiation skills

In addition to education, training, and experience, there is also a certification process for case managers. There is a national examination that must be completed successfully to qualify as a Certified Case Manager (CCM). The facility-based case manager will represent a single point of accountability to all those involved with caring for the subacute care patient, including the family or loved ones. A sample job description for a facility-based subacute care program case manager is presented in Exhibit 1.5.

Marketing Director

The subacute care marketing director is responsible for all marketing activities. The primary focus of this person is to increase census and improve payor mix through direct and indirect marketing activities. This individual should have skills combining

+ Marketing
+ Management
+ Organizational knowledge
+ Public relations
+ Healthcare
+ Negotiation

The marketing director should be energetic and interested in being a team player. Marketing has become recognized as a powerful force in the business of healthcare and, therefore, an entire chapter of this book is dedicated to marketing. A sample job description of a subacute care marketing director appears in Exhibit 1.6.

Director of Therapies

This position is usually performed by any of the following: physical therapist, occupational therapist, speech therapist, or in some cases a registered nurse with a specialty in rehabilitation. The director of therapies supervises therapy services for the subacute care program. The director of therapy services demonstrates knowledge of reimbursement methods and documentation in addition to experience in directing therapies in the healthcare setting. This individual is a key player in any subacute care program and especially one with a focus on rehabilitation. The director of therapies must be flexible, a good communicator, and actively interested in the success of the subacute care program.

Many other team members are important to the subacute care program, such as

+ Office manager
+ Respiratory therapist (RT)

- Physical therapist (PT)
- Occupational therapist (OT)
- Speech and language pathologist (SLP)
- Recreational therapist
- Social worker (SW)
- Registered dietitian (RD)
- Physical therapy assistant
- Certified occupational therapy assistant (COTA)
- Certified nursing assistant (CNA)
- Psychologist
- Neuropsychologist
- Admissions coordinator
- Receptionist
- Secretary
- Comptroller or financial director
- Billing and collection personnel
- Consulting physicians
- Maintenance
- Engineering
- Housekeeping

Patients/Family Members

The most important subacute care team member is the patient and the family or loved ones. Patients and their families become integral members of the team, and they are encouraged to actively participate in the program. The patients and their families are invited to case conferences and become part of the decision-making process. Subacute care programs emphasize education and training for patients and those persons who are involved in caring for them. The attitude and desire of the patient to return to the community is a significant factor that can make the difference between failure or successful outcomes.

Patients and their support system can become the program's best referral source or its worst marketing nightmare. The testimonials regarding their experiences with the subacute care program can be a powerful source of public relations.

SUMMARY

Subacute care evolved as a response to many needs related to the economics of providing healthcare. This industry is being touted by Wall Street as one of the most attractive niche markets for healthcare providers today. Subacute care programs are as good as the healthcare professionals that develop and implement them. These programs are a proactive response to the challenges that are presented because of evolving healthcare needs. The subsequent chapters are meant to provide insight into creating and administering successful subacute care programs.

APPENDIX

The following pages contain examples of job descriptions for some subacute care personnel. They are not meant to be complete but instead provide a foundation. Each subacute care organization should tailor these descriptions and position titles, and those for all other personnel, to meet the needs specific to the program, the services offered, and the patients who are admitted.

EXHIBIT 1.3
SAMPLE JOB DESCRIPTION FOR A MEDICAL DIRECTOR
OF A SUBACUTE CARE PROGRAM

Subacute Care Program
Medical Director Job Description

Reporting Relationship:

Accountable to facility administrator.

Qualifications:

The Medical Director will be a board certified, State Licensed Medical Doctor in the field most closely associated with the needs of the specific program.

Position Requirements:

The successful candidate must demonstrate a willingness to work with others in an interdisciplinary role. Must have a current, valid, and unrestricted license to practice medicine.

Job Summary:

The Medical Director will be responsible for no less than weekly consultations on each subacute care patient. The Medical Director will interface with other members of the interdisciplinary team and provide interaction with the family and education. They will also act as a liaison to the healthcare community.

Duties of the Director:

The subacute care program Medical Director shall perform services and functions that are in accordance with all applicable requirements

---

EXHIBIT 1.3
(CONTINUED)

---

of federal, state, local and/or facility laws, rules, and regulations. Such services shall include but not be limited to:

1. Conduct case reviews with subacute care program Case Manager to review and make recommendations for ongoing care as well as utilization review and discharge planning.
2. Available for education and inservice training of staff as needed.
3. Available to assist on preadmission assessment of potential subacute care program users to determine facility's ability to provide appropriate medical care.
4. Act as liaison with subacute care program consulting physicians and to medical community.
5. Assist in the development and approval of policies and procedures.
6. Attends and participates in case conferences as requested.
7. Participates in efforts to monitor and improve the quality of care in the subacute care program.
8. Participates as physician advisor to the facility interdisciplinary team and other committees as appropriate.

The subacute care program Medical Director:

- Must demonstrate knowledge and experience in specialized area
- Must have a license to prescribe controlled substances
- Must have a valid Medicare Provider Number
- Must carry professional liability insurance by an insurance company licensed in the same state as the subacute care program and be able to provide a current certificate of insurance
- Shall supply a resume or C.V. that includes but is not limited to proof of license, registrations, certificates, formal training, diplomas, and professional memberships
- Must be a staff member in good standing at the acute care facility that the program has a relationship with
- Must be willing to undergo a formal credentialing process and annual review
- Must supply information regarding any malpractice claims

EXHIBIT 1.4
SAMPLE JOB DESCRIPTION FOR A CLINICAL NURSE SPECIALIST
OF A SUBACUTE CARE PROGRAM

Subacute Care Program
Clinical Nurse Specialist Job Description

Reporting Relationship:

Accountable to the Director of Nursing.

Qualifications:

Required

Current licensure as a Registered Nurse, administrative and organizational skills. Able to execute sound judgment in patient care and staff development. Ability to implement interventions for patient care based on standards of nursing practice and principles of nursing care. Demonstrates supervisory management and leadership skills.

Desired

Minimum three years experience in the subacute or acute care relevant field, i.e., oncology, pulmonary, wound management, medical and surgical, infectious disease, rehabilitation, orthopedic care, pain management. Two years experience in supervision.

Job Summary:

Under the direction of the DON and supervision of the Administrator, the Clinical Nurse Specialist is responsible for the provision of 24-hour patient care to meet nursing needs of patients that include spiritual, emotional, cultural, social, and restorative needs and ensures that there are care policies and procedures designated to meet those needs.

In coordination with the Director of Nursing, ensures proper coordination of staffing, and scheduling of nursing personnel for assignment of duties that are consistent with their training and educational experience based on acuity within the subacute care program. The Clinical Nurse Specialist is responsible, in coordination with the Director of Nursing, for staff development, maintenance of nursing service objectives, and standards of nursing practice and has 24-hour responsibility for care management and staff development to meet care needs within the subacute care program.

EXHIBIT 1.4
(CONTINUED)

Job Responsibilities:

Any and all job related duties and responsibilities as directed by management including:

Clinical Responsibilities

Coordinates 24-hour patient care management within the subacute care program. Accountable for the clinical management of patients that includes but is not limited to:

1. Implementing interventions as indicated by the patients' plan of care or as ordered by the physician.
2. Ensuring 24-hour provision of services within federal, state, and company requirements.
3. Communicating with physician and interdisciplinary team as indicated by patients' needs or status.
4. Coordinating clinical services with other patient services.
5. Assessing and evaluating patients' change of condition and instituting appropriate interventions.
6. Initiating emergency measures according to facility's policy and within standards of nursing practice.
7. Making daily rounds to see all patients to ascertain condition, comfort, and level of care.
8. Assuming nursing responsibilities for minimum data set (MDS) of subacute patients.
9. Maintaining standards of services within the expectation of The Nursing Practice Act and the profession standards of practice.
10. Recommending number and level of staff for care to be staffed within the subacute care program.
11. Evaluating and identifying personnel for development and/or educational needs.
12. Monitoring personnel for competency in clinical skills.
13. Performing patient assessments as appropriate.
14. Initiating plans of care to promote desired patients' outcomes.
15. Monitoring personnel compliance to policies and procedures.
16. Initiating procedures to protect patients from injury and infection.

EXHIBIT 1.4
(CONTINUED)

---

17. Safeguarding the patients' privacy and property.
18. Coordinating admission of new patients.

Administrative Responsibilities

Develops and maintains clinical service objectives including but not limited to the following:

1. Maintaining standards of clinical practice.
2. Ensuring that only those patients for whom services can be provided are admitted to the facility.
3. Serving only one facility in the capacity of a clinical director.
4. Ensuring that the clinical function and training of the staff are met.
5. Ensuring that all documentation necessary for quality patient's care is maintained according to federal, state, and company policy.
6. Assisting in development and maintenance of budget for subacute care services as defined by facility budget.
7. Supporting the overall goals of the facility and the company.
8. Maintaining nursing policy and procedure manuals.
9. Ensuring exchange of essential information necessary for the provision of quality patient services.
10. Developing and maintaining written task guidelines for all nursing personnel.
11. Developing nursing policies and procedures as needed and in compliance with corporate standards.
12. Attending and participating in meetings and inservices as required.
13. Participation and support in subacute care marketing efforts.
14. Other responsibilities as delegated by administration.

EXHIBIT 1.4
(CONTINUED)

Supervisory Responsibilities

Responsible for supervision of all nursing personnel reporting to the subacute care program either by direct supervision or delegation, which includes but is not limited to the following:

1. Supervising all nursing personnel on their designated stations and responsibilities.
2. Making daily rounds, identifying problems, initiating corrective action, and monitoring follow-up.
3. Making decisions in emergency situations and being accountable for decision making.
4. Conducting systems reviews and staff meetings with clinical personnel to provide input and feedback on progress of assigned responsibilities.
5. Establishing department work goals coordinating delivery of care through supervision of nursing personnel.
6. Attends all care plan meetings to assess the total patient's needs.
7. Reviewing daily assignment with Director of Nursing, nursing supervisor, and/or charge nurses.
8. Monitoring nursing personnel performance.
9. Providing counseling and disciplinary action to all levels of nursing personnel as situation warrants. Making recommendations to Administrator and Director of Nursing if further disciplinary action is warranted.
10. Evaluating nursing personnel performance annually.

Consumer Service Responsibilities

1. Ensures that all patient/consumer rights are protected.
2. Adheres to company standards for resolving consumer concerns/complaints.
3. Participates in consumer education and training.
4. Projects professional image to consumers through dress, behavior, and speech.

EXHIBIT 1.5
SAMPLE JOB DESCRIPTION FOR A CASE MANAGER OF A SUBACUTE CASE PROGRAM

Subacute Care Program
Case Manager Job Description

Reporting Relationship:

Accountable to facility administrator.

Qualifications:

RN and/or CCM preferred. One to three years experience in medical rehabilitation, subacute care, care of chronically ill or catastrophically injured. Familiarity with available resources. Must be experienced in market trends, relationship development, maintenance of key accounts, and case management/managed care.

Position Requirements:

The successful candidate should have a willingness to assume a leadership role. Will demonstrate accountability and skills in decision making, written and oral communication, and time management. Should possess the ability to perform onsite evaluations, assess for appropriateness of placements, and implement and drive interdisciplinary care plans. Responsible for collecting and reporting outcome data and responsible for timely case management reports.

Job Summary:

The case manager serves as a single point of accountability to the payor, patient, and/or family for the care being rendered; its outcomes; and its cost effectiveness. The case manager is responsible for the screening of potential candidates, the development and implementation of the interdisciplinary care plan, within the financial constraints of the reimbursement allowed by the patient's insurer or funding source. The case manager provides an ongoing evaluation to monitor effectiveness of services, and determines that the established goals are being met within the appropriate time frames. Thus, these findings are to be directed to the appropriate referral/payor source through written reports.

The case manager participates in: identifying the subacute care needs of the local community, identifying local managed care referral sources, networking at case management meetings, rehabilitation and similar professional associations, and maintaining and developing

EXHIBIT 1.5
(CONTINUED)

relationships with referring case managers and hospital discharge planners. The case manager should also be a resource to administration and marketing, and participate in the decision making for product line development and the means and need to develop that product line.

1. Serves as a single point of accountability to the payor, patient, and/or family for the care being rendered; its outcomes; and its cost effectiveness.

2. Responsible for the screening of potential candidates, the development and implementation of the interdisciplinary care plan, within the financial constraints of the reimbursement allowed by the patient's insurer or funding source.

3. Provides an ongoing evaluation to monitor effectiveness of services and determine that the established goals are being met within the appropriate time frames.

4. Performs preadmission evaluations in conjunction with clinical nurse specialist.

5. Ensures that the assessment is documented to meet the facility and program's admissions criteria.

6. Communicates pertinent information in regard to patient/family.

7. Refers within system if patient's condition does not meet facility's criteria.

8. Makes determination of benefits by verifying eligibility, coverage, limitations, exclusions, and parameters of contract. Documentation of insurance verification with additional direction to the office for billing instructions.

9. Develops relationships with referring case managers and hospital discharge planners and provides support to marketing director.

10. Coordinates weekly medical rounds (where appropriate).

11. Rounds on all subacute care patients daily.

12. Coordinates and participates in team conferences to ensure documentation of patient progress. Collect written and verbal data from the team to produce a written case summary sheet, status reports, and discharge summary.

13. Obtains admission authorization and establishes level of care for that patient with payor.

EXHIBIT 1.5
(CONTINUED)

---

14. Informs patient/family and referral source of admission decision.

15. Works with interdisciplinary team to develop and implement plan of care to meet patient/payor/program requirements.

16. Directs (by mail or fax) care plan to payor source (if required).

17. Responsible for cost breakdown on managed care admissions. Coordinates with resources in the system (bookkeeping, pharmacy, rehab, nursing, etc.) to place patient at the appropriate level and maximize reimbursement.

18. Interfaces to evaluate appropriateness of care and treatment plan to enhance cost effective outcomes.

19. Obtains authorization for orders, communicates the changes to treatment team, and negotiates modifications of the treatment plan to the payor.

20. Educates the team regarding payor's requirements.

21. Coordinates discharge planning with social worker, checks if the expectations of the family/patient are met, that appropriate family training and education occurs, and helps determine home health and medical equipment needs.

22. Provides follow-up on discharged patients to be shared with team to evaluate the effectiveness of care plan, appropriateness of discharge planning efforts, and customer satisfaction. Follow-up will occur at a minimum after one month, six months, and one year post discharge.

23. Documents subacute care activity via current activity form. This is a means to obtain statistics for outcome studies, cost analysis, case studies, product line development, etc.

24. Performs home evaluations when appropriate.

25. Demonstrates accountability and skills in decision making, written and oral documentation, and time management.

26. Participates in tours with families, payors, physicians, etc.

27. Provides clinical assistance in the development of collateral marketing materials.

28. Provides backup to admissions coordinator in their absence.

29. Presents to external case managers, discharge planners, worker's comp, groups, etc.

EXHIBIT 1.6
SAMPLE JOB DESCRIPTION FOR A MARKETING DIRECTOR
OF A SUBACUTE CARE PROGRAM

Subacute Care
Marketing Director Job Description

Reporting Relationship:

Accountable to facility administrator.

Qualifications:

1. Education, experience, and training specific to managing marketing activities.
2. Experience in marketing healthcare services.
3. Ability to work effectively with all members of an interdisciplinary team as well as referral sources.
4. Verbal and written communication skills to perform tasks required.

Position Requirements:

1. Develops and implements facility marketing plan.
2. Ensures appropriate marketing systems (i.e., telephone, inquiry logs, complaints, physician relations, etc.) are functional within facility.
3. Develops relationships and conducts calls with key referral sources to initiate/increase inquiries and admissions to facility.
4. Facilitates marketing team meetings and projects to improve internal customer relations.
5. Develops and implements tracking system to monitor results of marketing efforts.
6. Identifies barriers prohibiting the achievement of budgeted census and mix goals.
7. Completes monthly marketing report.
8. Achieves facility census goals based on budgeted payor mix of Medicare, Private, Managed Care, Medicaid, and others.
9. Works on special assignments as requested.
10. Attends appropriate interdisciplinary meetings.
11. Initiates and negotiates relationships with payors.

EXHIBIT 1.6
(CONTINUED)

---

12. Manages and directs the marketing personnel.
13. Evaluates and reviews marketing efforts both internal and external.
14. Encourages the goals and objectives of the program and the parent organization.
15. Performs other duties as assigned by management.

CHAPTER 2

Developing the Subacute Program

*"Just think about what is going on around us.
That you and I should be living at such a time."*

— Dr. Zhivago

Regardless of whether a subacute care program is developed within an acute care hospital or within a nursing home, many of the same development issues arise. This chapter helps readers to gain insight about the planning elements necessary to successful development. The differences and similarities between a hospital location and a nursing home location are also addressed.

SUBACUTE CARE DEMAND FEASIBILITY

As discussed in the other chapters of this book, the goal of subacute care is to provide appropriate care to the patient population that no longer requires the intensity of inpatient acute care, yet for various reasons may still be unable to be discharged to the home. This may be true either because lesser levels of care, such as an outpatient or home health program, may not provide the intensity of nursing support or therapy required. The patient may not have the social support at home immediately available to provide monitoring and management. As important, these patients usually cannot be discharged to traditional nursing home providers, which typically offer programs targeting lower levels of skilled care. This is because traditional SNFs do not provide the nursing care, therapy support, or other ancillary services often required by the subacute care patient population.

To determine the feasibility of developing a subacute care program, two key elements must be considered. Each potential provider should first identify the patient population that will likely be referred and admitted to the subacute care program, then estimate the number of beds that will be required by this population. This effort will vary depending on the type of program being considered: either a hospital-based subacute care program or one affiliated with a freestanding nursing home. In general, a proposed hospital-based subacute care program will be the easier of the two alternatives to assess. This is because an acute care hospital (or acute rehabilitation hospital) presumably has an existing internal referral base from which to generate subacute care referrals and admissions.

A freestanding nursing home, on the other hand, must have external referral relationships with one or more local acute care hospitals or payors, or must have targeted specific physicians in the community to generate subacute care referrals. The hospital-based subacute care program will be discussed first.

HOSPITAL-BASED PROGRAM

As with developing any new clinical or diagnostic program, the hospital assessing the feasibility of implementing a subacute care program should first identify the potential patient population likely to benefit from and be referred to a subacute care program. This would include patients who may be categorized as requiring

EXHIBIT 2.1
SUBACUTE CARE REHABILITATION AND
SUBACUTE CARE MEDICAL PROGRAM DISTINCTIONS

Subacute Care Program	Nursing Care	Ancillary Use	Common Diagnosis	ALOS
Subacute Care Rehabilitation	4.5–5.5 hours per day	High use of PT, OT, and speech therapy	Orthopedics, neurology	7–21 days
Subacute Care Medical	4.5–8 hours per day	High use of respiratory, lab, pharmacy, and medical supplies	Cardiology, pulmonology, oncology, renal disease, postsurgical	7–21 days

either subacute rehabilitation or subacute medical care. Exhibit 2.1 summarizes these two subacute care patient classifications.

Although there may be considerable overlap between both patient categories and subacute care program types, certain analyses may be completed to identify the internal patient population. First, planners evaluate historical referrals to skilled nursing facilities, including the hospital-based skilled nursing unit if appropriate, and determine the number of referred patients by diagnostic category. Then, if the utilization review department maintains thorough patient management information, identify by clinical diagnosis the number of patients on skilled nursing facility waiting lists, and the associated patient days.

Planners must exercise caution during this phase of the analysis, however. Each of these reviews will only provide historical discharge and referral data, so this information should only be considered as an absolute baseline for a hospital's subacute care potential. These data will almost certainly significantly underestimate a hospital's true subacute care potential, because the historical referral patterns only reflect the utilization of traditional skilled care. These patterns do not capture all the patient classifications that might appropriately be managed in a subacute care program.

To better understand the hospital's entire subacute care potential, this preliminary analysis should be augmented with a detailed analysis of the hospital's acute care patient population. The most critical element in this process is identifying the patients who might appropriately be referred to subacute care, regardless of the historical treatment pattern. The most common approach to this phase of the analysis is first identifying the ICD-9 codes or diagnostic related groups (DRGs) most likely to generate subacute care referrals. Various research efforts have identified the diagnoses most likely to generate subacute care referrals. These typically fall into the following diagnostic categories:

- Cardiology
- General surgery
- Neurology
- Oncology
- Orthopedics
- Pulmonology
- Renal disease

These codes and diagnoses may be grouped according to patients who might be managed most appropriately in a subacute care rehabilitation program, and those who might benefit more from a subacute care medical model of care. Each of these patient groups should then be subdivided into patient age ranges;

specifically, each diagnostic group should be divided into those patients age 65 or older, and those patients younger than age 65.

Once assessment team members have identified by age the actual number of patients with each relevant discharge diagnosis, the team can apply the subacute care referral rate to each diagnostic category to estimate the potential number of subacute care admissions by category. The subacute care referral rates ranges from 25% to 75% of the total discharges by diagnosis, depending on the diagnosis. As would be expected, presently the referral rate for each category is significantly lower for the under 65 age category than the 65 and older category. In general, the younger population recovers much faster than the elderly population and is usually discharged more quickly to home, home health, or outpatient services.

The following steps summarize how to identify potential internal subacute care referrals for hospital-based units:

- ◆ Evaluate historical referrals to skilled nursing facilities.

- ◆ Determine number of referred patients by diagnostic category.

- ◆ Identify number of patients on skilled nursing waiting lists and associated patient days by diagnostic category.

- ◆ Identify ICD-9 or DRG codes most likely to generate subacute care referrals.

- ◆ Group codes into subacute care rehabilitation and subacute care medical patients.

- ◆ Subdivide patient groups into under 65 and 65+ age groups.

- ◆ Apply subacute care referral rates to each diagnostic category to estimate potential admissions.

Additional steps in the process of identifying potential internal subacute care referrals must be taken by both acute rehabilitation hospitals and/or units. If the facility considering startup of a subacute care program is either an acute rehabilitation hospital or an acute care hospital with an acute rehabilitation unit, three additional factors should be assessed. First and most important are the acute rehabilitation referrals and denials for admission. Denials are commonly issued because of the patient's failure to meet Medicare admission criteria, such as ability to tolerate three or more hours of therapy per day. Frequently, however, these patients are perfect candidates for subacute care, which does not have the same admission constraints as acute rehabilitation. Most acute rehabilitation providers have found that approximately 40% to 60% of all patients receiv-ing acute rehabilitation denials are candidates for subacute care rehabilitation.

The second adjustment for the acute rehabilitation hospital/unit is esti-mating the current acute rehabilitation patient population most likely to be forced by insurers and other payors into a lower-cost setting. Most often, this estimate will include the bulk of the acute rehabilitation orthopedic patients who are currently managed in subacute care rehabilitation in many markets, and 20% to 30% of other general rehabilitation diagnoses such as stroke. This, of course, will reduce the acute rehabilitation census proportionally and should be a factor in the project assessment.

Finally, the acute rehabilitation hospital or unit should consider its dis-charges to skilled care following the acute rehabilitation phase. This often represents 10% to 20% of the total acute rehabilitation patient population, and is an internal source for the hospital's own subacute care/skilled care unit.

Here, then, are the additional steps for rehabilitation hospitals and hospitals with acute rehabilitation units:

- Examine acute rehabilitation denials for subacute care rehabilita-tion appropriateness.

- Estimate current acute rehabilitation patients likely to be encour-aged by payors to use a lower-cost setting.

- Consider current rehabilitation discharges to skilled care.

Once the potential internal acute care and, if appropriate, acute rehabili-tation referrals, have been identified, the next critical step is estimating the number of subacute care bed needs. To do this, planners estimate average lengths of stays for the various subacute care patient populations and multiply each estimate by the number of potential hospital referrals by diagnostic mix. The average subacute care patient's length of stay will vary by diagnostic mix, although most diagnoses for those 65 or older will be between 7 and 21 days. Consistent with the lower referral rate, patients who are 64 or younger typically have shorter lengths of stay in subacute care programs. Generally the younger population's length of stay is half that of the older patients.

After the potential patients have been identified and the projected patient days have been estimated, the internal bed need is simply a mathematical calculation. First, divide projected days by 365 to determine the projected average daily census for the subacute care program. Next, divide the average daily census projection by the target occupancy rate to estimate the number of beds. Typically, because of the relatively rapid turnover rate, an 80% to 85% occupancy rate is targeted.

Thus, the steps in estimating the subacute care program's bed needs are:

1. Estimate the average lengths of stay (ALOS) for each subacute care patient population.

2. Multiply the ALOS estimates by the number of potential hospital referrals by diagnostic mix to project the total patient days.

3. Divided the total patient days in the program by 365 to determine the average daily census.

This planning process should result in a reasonable estimate of the expected number of internal subacute care referrals and resulting bed need, based on current subacute care referral rates and bed utilization. Any hospital's specific projections, however, should be flexible enough to allow for adjustments to be driven by characteristics unique to that particular hospital or market.

Factors that might influence the utilization projection either positively or negatively include:

- Conservative/aggressive medical staff practice patterns

- Internal clinical pathways for specific diagnoses which include levels of care other than acute care

- Internal case management systems

- Managed care environment

- Specific managed care contracts

Additionally, adjustments to the subacute care bed requirements are frequently made to capture those patients who likely fall into a wide variety of diagnoses and who have exceptional lengths of stay, usually greater than 25 to 30 days. These patients usually have comorbidities and would commonly be categorized as subacute medical patients. In some hospitals this patient population may be as large as the previously identified subacute care population. If the long-term patient population is increased, the hospital evaluating subacute care may want to consider the development of a long-term hospital.

Although commonly referred to as a *subacute* care provider, a long-term hospital actually has an *acute* care license and has a total hospital average length of stay greater than 25 days. A hospital meeting this criterion is exempt from the Medicare Prospective Payment System and is reimbursed under a cost-based mechanism similar to that for acute rehabilitation providers. Although this is more problematic than developing a subacute care program through hospital-based skilled nursing beds, for some large acute care providers or for a network of acute care hospitals this may be an attractive development alternative.

In addition to projecting the utilization of a hospital-based subacute care program, a hospital evaluating the development of such a program should also estimate the impact the unit would have on acute care utilization. Most certainly the successful implementation of a subacute care program will decrease the patient days and average daily census of the acute care patient population. In fact, because of the Prospective Payment System and other fixed-payment reimbursement systems, the reduction of acute care patient days is a primary goal.

In order to estimate the impact on acute care utilization, hospitals should review the estimated subacute care referrals and admissions by patient diagnosis, and estimate how the acute average length of stay for each of these patient types will be reduced by the subacute care program. In general, hospitals can expect that the subacute care rehabilitation and subacute care medical patients will have an acute length of stay that is approximately three to six days less than that occurring before the subacute care program became operational. The long-term subacute care patients discussed earlier, while smaller in number, will realize more dramatic decreases in length of stay. It is possible to reduce the acute length of stay for many of these patients by 30 days or more, depending on the support available on the subacute care program.

The final element of demand feasibility for a hospital-based subacute care program is screening these admission and patient length of stay projections against qualitative information collected through internal interviews. Physicians from multiple specialties; clinical managers; and administrative representatives from operations, nursing, and finance are among those who should be inter-viewed to ensure all assumptions are reasonable. Subacute care referral rates used to estimate internal demand are benchmarks used by many hospitals. However, each hospital is unique and will have a specific environment that might influence the development of subacute care.

As with any new-product analysis, in addition to the demand feasibility data, it is also imperative to complete a financial feasibility study. This should include estimating program costs, charges, and reimbursement. Medicare reim-bursement must be estimated appropriately during the first three years of the program's operation, which are cost-based, and during all successive years, which apply a routine limit. Subacute care financial feasibility is discussed in detail in Chapter 5.

FREESTANDING NURSING HOME FACILITY

Freestanding nursing homes considering the feasibility of developing a subacute care program will want to follow many of the same steps outlined for acute care hospitals. These facilities do not have an internal acute care population from which to draw referrals, so the identification of potential subacute care patients will be somewhat more complex. Ideally, a freestanding nursing home would have one or more acute hospital partners from which to draw subacute care referrals.

Hospitals become interested in jointly developing subacute care programs with nursing homes for several reasons. The most frequent reason is that a hospital may be prevented by Certificate-of-Need (CON) rules from obtaining licensed skilled nursing beds. Additionally, some hospitals prefer not to enter into subacute care because it most often must be managed under the many constraints of skilled care licensure requirements, which are significantly different from acute care licensure requirements.

If a freestanding nursing home has identified potential acute care partners, the nursing home administrative staff will want to follow the same process as that outlined for acute care hospitals in order to estimate the potential subacute care referrals from the acute care patient population. Experience has shown, however, that nursing homes should develop more conservative estimates for referrals than hospital-based subacute care programs do. Physicians frequently are reluctant to discharge patients to alternative programs if the programs are not located physically within the hospital or in close proximity.

Some freestanding nursing homes have entered the subacute care business without acute care hospitals as partners, but rather with specific physicians as formal or informal partners. If a nursing home is considering this alternative, an assessment similar to that of the acute care hospital must still be completed, but with some practical modifications. Rather than examining acute care discharges from the hospital, the nursing home should evaluate all acute care discharges and discharge practice patterns for the specific physician(s). If committed to referring all appropriate patients, some active physicians in specialties such as orthopedics or neurology could refer enough subacute care patients to develop a viable subacute care program. As with the acute care hospital, the nursing home should evaluate the physician's patient population, examine the subacute care potential, and calculate the number of patient days and subacute care beds required. If this approach is taken, it is even more important to conduct multiple physician interviews. These individuals will likely control the lion's share of potential referrals and subsequently determine the fate of the proposed program.

If the freestanding nursing home has neither an acute care hospital nor any specific physicians as partners in the subacute care project, or a managed care company, then an external assessment of the service area will be required to determine the current demand and supply for subacute care. Two approaches of measuring demand can be used. The first approach is a population-driven methodology of determining community demand. The second is based on acute discharge data that might be available through either the area hospital association or private vendors.

The population-based methodology is based on a nationally estimated subacute care bed-to-population ratio of approximately 0.5 to 0.6 subacute care beds per 1,000 population. This, of course, represents the estimated bed need for a service area with an age distribution consistent with that of the national population projections. A freestanding nursing home that estimates demand in this manner should adjust the bed need ratio to reflect the demographic profile of its specific service area. If the community has an older or younger population profile than the national distribution, the bed need ratio will need to be increased or decreased proportionally.

The acute discharge data method may be employed if such data for the service area are available. From these data, subacute care bed need may be determined in the manner presented for acute care hospitals. First, identify the specific DRG and ICD-9 codes likely to generate subacute care referrals. Then multiply the corresponding cases by the appropriate subacute care referral rates. Patient days and bed need may also be estimated using a process similar to that for the acute hospital analysis. This will allow the nursing home to identify subacute care bed need by ZIP code, county, and/or acute care hospital.

The following are steps in forecasting external demand for a subacute care program:
1. Apply a .5 to .6 beds/1,000 population ratio to service area population.
2. Adjust ratio for service area demographic variations from national distribution.
3. Identify existing subacute care providers/beds in market.
4. Subtract existing beds from estimated demand to determine net need.

Once the freestanding nursing home planners complete their estimate of the service area bed need, they will then identify the existing providers and number of subacute care beds to distinguish any true unmet need. The existing providers may be identified through interviews with local physicians, social workers, and discharge planners, and also through local advertisements and the Yellow Pages. A telephone call to these potential providers should reveal specifics about their programs. Any subacute care provider should offer to a community inquiry the following information, which will offer insight into the program:

- ◆ The number of beds dedicated to subacute care
- ◆ Whether the subacute care beds are clustered together on a unit or spread throughout the nursing home
- ◆ The types and number of therapists the nursing home has on staff
- ◆ Any special subacute care programs offered

Additionally, third-party information can frequently be collected from state planning agencies. Subacute care is not a separate licensure category in most states; therefore, some educated guesswork is required.

Information reported on annual nursing home questionnaires (which will vary by state) may offer several indicators of the level of care provided by individual facilities. These include staffing levels, Medicare payment, patient charges, and discharge rates to home. All of this information enables a planner to determine whether a particular nursing home is truly providing subacute care, or, as is often the case, has simply renamed its current services without implementing any program-wide changes.

After both the demand and current supply have been estimated, the freestanding nursing home planners should be able to determine the feasibility of subacute care development from a service-area need perspective. Assuming that the unmet need is positive and the subsequent financial analysis is positive, it is critical that the freestanding nursing home develop an educational and marketing plan that targets specific markets from which to generate substantial referrals. Marketing subacute care is discussed in more detail in Chapter 9. This plan will ensure that the external referral sources continue to refer a population sufficient to meet the program goals.

SUBACUTE CARE PROGRAM SIZE AND LOCATION

If, based on the demand and financial feasibility, a hospital or freestanding nursing home planning team decides to develop a subacute care program, both the size and location of the unit will depend on several factors. Of paramount importance is the bed need to service subacute care patients. As with any inpatient program, the greatest consumption of actual space will be patient rooms, so the actual number of beds and desired private to semiprivate room ratio will determine most of the square footage requirements.

Second to bed need will be the program's administrative and ancillary support requirements. Office space for a medical director, program director, and nurse manager will be required, as well as space for social workers, family conferences, and staff support such as a secretary. Frequently, the support space will be 20% to 30% as large as the patient care space.

In addition to bed need and support space, because most subacute care programs are developed as licensed skilled nursing beds, the state licensure requirements for nursing beds will also dictate space requirements. Most often, state requirements for nursing home facilities are more demanding than those for acute care facilities. They include total square feet per bed, patient room distance from the nursing station, and availability of day rooms and common areas, hairdressing/barber shops, and dining rooms.

Finally, a new subacute care program should have adequate space for any unique programmatic elements of the service. For instance, if the program focus is subacute care rehabilitation, the provider will need to ensure that ample space is planned for therapy areas, so that most therapy can be delivered on the unit without transporting the patients throughout the facility.

Following are key factors determining subacute care program size and location:

- ◆ Number of beds
- ◆ Private/semiprivate room complement
- ◆ Administrative and ancillary support requirements
- ◆ State requirements for skilled nursing beds versus acute beds
- ◆ Adequate space for program emphasis elements (such as therapy space)

Most often, the actual location of the unit depends on space available at either the hospital or the freestanding nursing home. Location within the physical plant should be driven more by adequate space needs than by the need to be located close to any specific hospital/nursing home program or service. As with an acute unit, most ancillary services will be provided in the subacute care program.

SUBACUTE CARE CONSTRUCTION AND EQUIPMENT

Similar to unit location, the construction required for development of the subacute care program depends on space available within the hospital or nursing home, and the previous use of that area. If, for instance, the space targeted for the subacute care program was not previously used for patient care services (or if completely new construction is required), the facility should expect to incur relatively significant capital expenditures. If, however, there is ample support and ancillary space and the area previously was used for inpatient care, it is possible to develop a subacute care program with lower capital investment.

Regardless of previous uses of the targeted space, however, certain factors should be considered. High on this list is functionality of patient bathrooms, because this feature tends to be an expensive component of many conversion and renovation projects. Improving the functional level of patients and discharging approximately 80% to 90% to their homes is one of subacute care's major goals. Therefore, patients should be able to manage activities of daily living themselves, including using the restroom. Ideally, each patient bathroom may also have a shower. However, the patient population and state regulations will dictate that some/all of the patient bathrooms be handicapped- and wheelchair-accessible.

In addition to patient bathrooms, any construction and renovation should also incorporate features for the special needs of certain targeted patient populations. This may include piped-in oxygen for ventilator-dependent patients, isolation rooms, therapy space for rehabilitation, and any other special patient needs.

Consistent with the architectural guideline to address both form and function, cosmetically the subacute care program should be designed with the specific population in mind. This should influence color selection, use of textures, and design and use of furniture and adaptive equipment.

In a freestanding nursing home, cosmetics are likely to be more important than in a hospital-based subacute care program. The freestanding nursing home planners will need to differentiate the subacute care program more clearly from the rest of the nursing home in the minds of payors, physicians, and consumers. If possible, the unit should be distinct from the traditional skilled nursing unit. Even though they will be occupying beds most often licensed as skilled care, patients admitted to subacute care do not want to be admitted into a routine nursing home.

The basic equipment needs of the subacute care program may be similar to those of a skilled nursing program in terms of patient rooms and nursing services. Depending on the targeted patient population, the use of certain therapies and higher-technology equipment may be needed. This might include items such as occupational therapy testing and training equipment, or state-of-the-art monitoring and treatment equipment for the ventilator-dependent population. The key ingredient in planning for equipment within a subacute care program is consideration of the target patient population by clinical diagnosis and age range.

SUBACUTE CARE REGULATORY PROCESS

Depending on the hospital or freestanding nursing home's target subacute care market and the organization's current bed complement, the regulatory process may be either relatively easy to manage or one of the most difficult aspects of the subacute care development project. If the organization is targeting solely the commercial and/or non-Medicare patient population, in states that require Certificate-of-Need (CON) approval for skilled nursing beds, there are fewer obstacles for the organization to manage. Why? The answer is because technically subacute care may be provided in any type of licensed bed. However, for the Medicare patient population the facility will not be reimbursed beyond the DRG payment if the patient is not admitted into a licensed skilled care bed. If the target population is non-Medicare, the program may be developed for any bed license and the payment negotiated by the level of care provided instead of the type of bed.

In most markets today, however, a significant percentage of the subacute care patient population is covered by Medicare. Therefore, Medicare-certified skilled beds should compose a greater number of beds in the subacute care program. If the hospital or freestanding nursing home already has skilled nursing beds available for conversion within its bed license, the subacute care program may be developed without any need for CON approval.

If new beds will be required, a CON application must be filed with the appropriate state agency and an approval received before the program becomes operational.

Although the process varies among the states, in general the hospital or freestanding nursing home must include in the CON documentation both substantiation of program need and financial feasibility. Usually the planning conducted to determine beds needed, program focus, and financial return can be modified to conform to application guidelines. The application should address the internal needs of the hospital/nursing home, difficulty in placing current subacute care level patients within the community, and a clear description of the difference between subacute care and traditional skilled care.

Subacute care is not specifically addressed in most states' CON standards and criteria for skilled nursing beds; therefore, the state planning agency may not look favorably on allocating community skilled nursing beds from an officially designated skilled nursing facility bed need to subacute care programs. However, many states are beginning to recognize the value of converting underutilized acute hospital beds for this purpose, as well as providing subacute care in the less expensive nursing home setting. Both hospitals and freestanding nursing homes therefore can often prevail over traditional nursing home providers in competitive review situations.

However, despite the need for subacute care services in a service area, some states have moratoriums or strict development guidelines for skilled care beds. If this is the circumstance, the program's focus may be redirected to any of several secondary options. These include purchasing existing or CON-approved beds, developing the subacute care program in partnership with an existing provider of skilled care, leasing skilled care beds within a nearby existing provider, allowing an existing skilled care provider to lease space and relocate skilled care beds into the facility, or developing a subacute care program exclusively for non-Medicare patients in other bed-license categories.

Following is a summary of the range of alternatives to seeking CON-approved skilled nursing bed approval:

- ◆ Purchase existing or CON-approved beds.
- ◆ Develop subacute care program in partnership with an existing skilled care provider.
- ◆ Lease skilled beds within an existing provider's facility.
- ◆ Lease internal space to an existing skilled provider.
- ◆ Develop a non-Medicare subacute care program under another bed license category.

In any case, as a facility plans a subacute care program, the organization should be aware of the state's review process and time schedules. In many states this process routinely takes from 3 to 12 months.

In addition to CON requirements in most states, all states have some nursing home licensure requirements to which the facility must adhere if nursing home beds are used. These requirements address space, location, and staff support. A state surveyor usually visits the unit before its opening to ensure that licensure standards are met. In addition, some state licensing departments require submission of facility blueprints or other information during the development process to monitor progress. Also, in most states the licensure department has the authority from the Healthcare Financing Administration to certify adherence to Medicare as well as state guidelines.

Finally, in addition to CON approval and licensure requirements, many providers will want to consider additional industry certifications. These certifications usually are designed to measure program quality and patient satisfaction, and might be used by the facility in several ways:

- ◆ Approval by a third-party licensing or accreditation body ensures the provider that certain standards are being met and serves as a measurement guideline.
- ◆ Receiving such a licensure or accreditation may be used as a marketing tool to distinguish the subacute care provider from others positioned in the market as offering similar care.

◆ Most of these external review agencies, which include the Joint Commission on Accreditation of Healthcare Organizations (JCAHO) and the Commission for Accreditation of Rehabilitation Facilities (CARF), usually update standards and provide new guidelines so that advances in the industry are communicated and implemented. These two agencies are discussed in further detail in Chapter 11.

DEVELOPING POLICIES AND PROCEDURES

Policy and procedure development depends on a number of factors, including Medicare certification requirements, target patient population and program strategy, program model of care (that is, subacute medical or rehabilitation), and accreditation requirements. Many individuals fall into the trap of thinking that written policies and procedures are the hoops to be jumped through in order to meet certain industry requirements. However, if policy and procedure development is viewed more accurately as the mechanism through which much of staff education and training will occur, the end product will be far better developed and far more useful than if viewed in any other way. This is particularly true within the context of a subacute care program's development because, in most instances, all members of the development and treatment team are on a steep learning curve. Medicare certification requirements form the backbone of any policy and procedure manual for a subacute care program. These requirements were written for the traditional long-term care industry, so they pose problems for both the acute care and the long-term care provider.

For the acute care provider, development of policies and procedures aimed at meeting Medicare-certification requirements for skilled nursing is foreign territory indeed. Although nursing procedures from within the hospital may transfer to the needs of the subacute care program, the administrative policy requirements are significantly different. The acute care provider anticipates fairly sick patients will be transferred to the subacute care program, but hospital management must be cognizant of two facts if they are to succeed in their subacute care program development:

◆ Nursing homes are considered long-term residential facilities. The word *resident*—not *patient*—is used in the certification standards. Therefore, resident rights issues compose a significant segment of the requirements. Regardless of the acuity level of the "residents," all certification requirements must be met.

◆ Medicare views nursing home beds within a hospital no different from those in a freestanding nursing home. Therefore, contractual arrangements with ancillary departments of the hospital do not exempt those departments from the same requirements as those within a freestanding nursing home.

Thus, the development of policies and procedures for a hospital-based subacute care program requires significant coordination and cooperation among the members of the hospital's management team if Medicare certification is to be achieved. Hospitals entering the subacute care arena for the first time should develop an implementation team that includes members from all departments with even the slightest responsibility to the unit. At a minimum, the team should consist of the members as noted in Exhibit 2.2.

EXHIBIT 2.2
IMPLEMENTATION TEAM PROFILE

Team	To Be Included
Core Team	Administrator or manager, medical director, nursing, rehabilitation therapies, social work, respiratory therapy, finance/reimbursement, case management, public relations, marketing, dietary, medical records, and others depending on hospital department makeup
Ad-Hoc Team	Laboratory, radiology, central supply, pharmacy, planning, education, and others depending upon target patient population

---

Medicare certification criteria form the foundation for the unit, so unless managed care is the only target, attention must be given to developing policies and procedures that meet both the criteria and needs of the target population. Conflicts between the two goals commonly emerge. These usually occur when the cost of meeting patient needs is compounded by Medicare requirements for meeting resident needs. However, as a general rule, management should err on the side of meeting Medicare requirements while still trying to achieve stated patient goals in the least costly manner.

Additionally, because the Medicare requirements are foreign to most acute care providers, it is recommended that an implementation plan and checklist be developed. Exhibit 2.3 gives an example of a policy development checklist.

Once policies and procedures are completed, a final review against the Medicare Manual and the written documents should be completed to ensure that all elements have been addressed.

EXHIBIT 2.3
SAMPLE OF A POLICY DEVELOPMENT CHECKLIST

POLICY DEVELOPMENT CHECKLIST

Tab #	Policy	Interpretation	Date Due	Date Done	Date Rev.	Who Resp.

For the freestanding nursing home, the elements of policy and procedure development under the Medicare program are somewhat reversed from that of the hospital. Although the hospital must switch to a "residential" mind set as administrators develop policies and procedures, the nursing home must begin to change its thinking to acute care. That is, care must be taken in developing policies that do not undermine Medicare rules and regulations but that also can accommodate both high acuity needs and rapid turnover. The average subacute care length of stay is less than 20 days. Therefore, management and clinical team members must think through the best mechanism for handling issues associated with short-term residents (relative to the typical nursing home resident) and document policies that will serve as true resources to staff.

Like the acute care hospital, freestanding nursing homes should develop a core development team as well as identify ad-hoc members who would be involved on an as-needed basis. Additionally, an implementation plan should be developed that identifies the issues to be addressed over and above the existing Medicare requirements, and the mechanisms and responsible parties for addressing them.

With Medicare policy requirements completed, it is now time to tackle policies that will address the strategic issues that triggered the development of subacute care in the first place. From a policy and procedure standpoint, the issues critical to achieving the program's strategic goals usually are embedded in model-of-care and admission and discharge policies.

Regarding models of care, if thoroughly completed, the planning process accurately identified the target patient population for the program. The types of residents (rehabilitation, respiratory, and so on) that the facility will serve tends to dictate the model of care to be developed, which in turn determines the policies and procedures that will govern the program. Typical models of care include subacute care rehabilitation and subacute care medically complex diagnoses. Subacute care medically complex diagnoses may include short- or long-term situations. Although rehabilitation therapies tend to be included in all of the models, the unit should be considered a rehabilitation unit when the primary reason for patient admission is functional improvement rather than medical recovery.

Admittedly, gray areas exist between models of care and policies and procedures, so development should address such issues as commingling patient types, programmatic structure, physician admitting policies, program description and direction, and program outcome expectations.

The patient population that will be admitted to the program should be identified early on. When no predetermined criterion offers any guidance, it is incumbent on management to define the target population through the planning process and to clearly articulate how patients will be evaluated for admission and discharge. Programs lacking clearly defined admission policies will quickly learn that the goals that precipitated strategy development have not been met. In fact, not only can the strategic vision for the program remain unrealized, but also the chance of developing a financial drain on the organization can occur.

Critical items for admission criteria:

- Target diagnosis
- Clinical stability
- Potential for additional acute care procedures
- Potential for medical or functional improvement
- Estimated length of stay relative to program goals
- Family support
- Discharge destination
- Patient's financial resources

Following development and sign-off of the admission criteria, it is essential to develop an admission process policy. The policy should define how patients will be evaluated for admission, who is responsible for the evaluation, and timeliness of reply to the referral source. The policy should reflect the authority of the individual making the admission decision. Federal requirements are stringent, requiring much attention to be paid to this element of policy and procedure development.

Finally, should a facility choose to participate in accreditation, another set of mandated policies and procedures applies. As mentioned earlier in this chapter, two organizations accredit subacute care programs: the Joint Commission on the Accreditation of Healthcare Organizations (JCAHO) and the Commission for Accreditation of Rehabilitation Facilities (CARF). The JCAHO standards apply to subacute care programs regardless of whether the program employs a medical or rehabilitative model of care. The CARF standards apply to subacute care rehabilitation programs.

For hospital-based subacute care programs, the JCAHO standards should pose little difficulty, because all hospitals that participate in the Medicare program must be JCAHO accredited. Although many of the subacute care standards are new to hospitals, familiarity with JCAHO standards should make appropriate policy and procedure development essentially an "add-on" to current manuals.

However, many freestanding nursing homes are not JCAHO or CARF accredited. For these facilities, development of policies and procedures to meet accreditation standards may be a formidable task. Workshops are usually offered for new applicants, and new subacute care providers should take advantage of such training prior to preparing for the all-important accreditation site visit.

Like JCAHO, CARF standards are familiar to hospitals with acute rehabilitation beds. Similar to JCAHO, CARF standards provide new concepts for providers that have not had organized rehabilitation programs in the past. In this regard, the acute care hospital and the nursing home provider may both require some training prior to the site visit. This is discussed in more detail in Chapter 11. Because any accreditation process is steeped in written policies and procedures, providers must take the time to work through their program development, document all of the appropriate policies, and review them against the standards. Policies should be written to reflect actual practice and evolve with program development so they remain dynamic tools for staff education and orientation.

STAFFING THE SUBACUTE CARE PROGRAM

Just as subacute care programs bridge the gap between acute care and traditional nursing home care, staff expertise should span that range. Regardless of whether the unit is in an acute care hospital or nursing home, a blending of acute care and long-term care skill is critical to the program's success. Staff credentials should reflect the needs of the target patient population. That is, if the unit is predominantly a subacute care rehabilitation program, care should be taken to staff the nursing component with nurses trained as rehabilitation nurses. Therapy staff should also reflect the patients' needs. This staff should be assigned consistently to the subacute care program and be part of the program's structure. Similarly, subacute care medical programs should be staffed commensurate with the needs of the patient population they serve. If the unit is targeting postsurgical, cardiac, or respiratory needs, for example, staff should have some critical care experience.

Along with staff experience and expertise, direct-care hours should be based on the acuity needs of the patient. Although the average number of direct nursing hours for a subacute care program ranges from 4.5 to 5.5 hours, programs with higher acuity patients and/or ventilator-dependent patients may provide as high as 8 or 9 hours of direct care. However, a key difference between acute care and subacute care is the mix of nursing staff; acute care usually staffs with a higher professional mix (greater than 60%) than a subacute care program (less than 50%).

SUMMARY

For the appropriate patient population, subacute care has proven to be both an effective clinical and cost-efficient model of care. It is important, however, that any organization considering the development of a subacute care program follow a planning process that reduces the likelihood of encountering unforeseen obstacles and increases the chance of long-term success. A thorough assessment of the subacute care opportunities before implementation allows the organization to identify the potential patient population and dedicate the correct number of beds for subacute care. Further, such an assessment promotes design of the operational program to meet the needs of the patient, physician, and hospital or nursing home.

The net outcome of an effectively planned subacute care project should be a new product line that produces high patient and physician satisfaction, quality clinical outcomes, and positive financial experiences for the patient, payor, and provider. These outcomes can be only achieved, however, if the organization has made an early commitment to understanding the dynamics of this evolving level of care and to following a sound planning process.

PART II

◆

SUBACUTE CARE
MODEL SAMPLES

CHAPTER 3

A Respiratory Care Model

"So much to discuss, so little time."
— Dr. No

Respiratory/ventilator patient care units were some of the earliest examples of nontraditional subacute care program settings in the country. The typical long-term ventilator patient profile perfectly matches the design and goals of subacute care, because the patient has complex medical needs, relative medical stability, substantial potential for rehabilitation, and mounting high costs to payors. Interestingly enough, patients with increased technological needs also pose more challenges to facilities wishing to develop a respiratory/ventilator program. Reimbursement, remodeling, equipment, staffing, and physician support are all crucial components of any subacute care program setting. Yet, they often can be exponentially more involved in programs where life support is a primary goal.

The first and often-overlooked place to start is a needs assessment. Many people in healthcare professions can offer information about the need for a specific type of patient care. Ventilator patients often stand out in the minds of discharge planners, case managers, and administrators as being patients who had amassed high daily charges, had little to no long-term reimbursement funding, and offered few if any placement alternatives. This is often translated into a significant community need although this may or may not be the case. It is easy to fall into the trap of believing that these patients require a high placement. Quantifying that perceived need is imperative.

There are many ways to perform the assessments. The first step is to do a feasibility study. For more details regarding feasibility and needs assessment, see Chapter 2. In determining ventilator patient census, you must understand how the long-term ventilator patient is defined by the acute care hospital. A typical definition is a patient who has been on ventilation for 21 consecutive days. However, some facilities may extend that time frame to include several months. It is important to establish that these facilities have a need for placement of these ventilator patients. Professionals will ultimately control referral patterns of patients who have been considered to be at the acute care level. Their buy-in of need, financial benefit to the community, and quality of medical care in a nontraditional setting is crucial.

Administration must understand the financial benefit to the facility. An acute care facility could benefit from early discharge of high-cost patients. It is important that a subacute care program be viewed as an adjunct to the referring hospital and not seen as a competitor.

Early discharge means more cost efficient utilization of the Diagnosis Related Group (DRG) payments from Medicare. In addition, an early discharge to a quality subacute care respiratory/ventilator program can elicit a positive response to rehabilitation and eventual discharge from the medical system. This results in fewer dollars spent per patient. Most hospitals in America are moving toward the lower cost, quality, and outcome-focused providers in their area. An agreement between the hospital and a subacute ventilator program offers significant potential for success.

Pulmonologists and other physicians should be involved in establishing the medical care delivered and what services will be available to their patients. A respiratory/ventilator program should employ a pulmonologist who is willing to be a medical director. In addition, the medical director must be willing to promote the unit within the community. He or she can in turn become a primary agent in marketing the idea of a subacute respiratory program to other physicians. Pulmonologists should also be involved in developing the policies and procedures for the unit.

Physician support is obviously important but can prove to be quite difficult to achieve because of perceptions of long-term care and reimbursement. Reimbursement policies for physicians making visits in long-term care facilities has been misunderstood. The Health Care Financing Administration (HCFA) recently increased the allowable number of visits and charges per month for long-term care, making it more attractive to see Medicare patients in a long-term care setting. In addition, a facility can directly reimburse a physician—particularly a specialist such as a pulmonologist—for medical director's fees and patient services.

Size and location of the program are important. A program that is small will not achieve the economies of scale necessary to offset high operational costs. A program that is too large may have problems with poor space utilization and negatively impact cost report data. The minimum number of beds recommended for a respiratory/ventilator program is eight. The average number of beds used for a respiratory/ventilator program is 11–20.

A respiratory care facility should be located in an area that can be isolated from the main patient population. Separate entrances and access to parking are preferred by some providers but not considered necessary by others. Significant storage, office, and patient space will be needed. Individually controlled heating and cooling are also necessary. Patients with chronic lung disease or pulmonary insufficiency can be sensitive to the room temperature. Rooms that are perceived to be too hot or cold will impede the weaning process for ventilator-dependent patients because patients may not be able to focus on their rehabilitation program.

The level of equipment and accessories specific to ventilator-dependent patients play an important role in determining room dimensions. In an average size semiprivate room, two ventilator patients and their accompanying array of equipment easily fill all available space. It is important to include enough space to provide patient care and to get patients in and out of their rooms. A minimum size for a semiprivate room should be 180 square feet. Ideally, rooms should be private because of the complications of cross contamination. Yet, this is not always acceptable from a cost analysis perspective.

Availability of shower facilities must also be considered when planners design a location. Facilities should have both shower and bath capabilities. Patients with decreased mental status may not easily be accommodated in a tub. Manufacturers make rolling bathtubs that allow staff members to place a patient inside and then roll them to a shower area.

ADMISSION CRITERIA

A program must have a clear focus regarding the criteria of patients it will and will not accept. These criteria are several factors in addition to community need. Following are some criteria that should be addressed when admitting patients with respiratory/ventilator needs to a subacute care program:

- Age
- Payor type
- Diagnosis
- Code status
- Rehabilitation potential
- Targeted length of stay
- Acuity

- Support system (family, loved ones, and so on)
- Equipment needs
- Infectious status and history
- Discharge alternatives

Subacute care is a newer concept to the general community. Education of the community as to what subacute care is and how your facility will interact with the current medical community is imperative. Clear admissions criteria along with an educational plan will assist your program in ensuring a consistent and appropriate referral pattern. Determining the patient focus early and being tenacious will set clear examples of types of the patients the program can and cannot care for. There is widespread need for respiratory/ventilator subacute care programs in many parts of America. It is simply a matter of determining what is right for the specific facility and the needs of the community.

Besides the issue of life support, infections are perhaps the most common denominator in long-term respiratory/ventilator patients. Due to the invasiveness of tracheotomies, accompanying tubing, and paraphernalia that are common with this level of care, infections are frequent and often the greatest barrier to patient rehabilitation. Specifically, resistant bacteria and the expense of treatment including pharmaceuticals are key areas to consider prior to opening a unit.

It should be established, prior to opening a respiratory/ventilator subacute care program, whether patients with a current or recent prior history of resistant bacteria will be admitted. Many ventilator patients have contagious bacterial infections. It is important to decide whether this will pose a significant problem with the current patient population, staff members, and state accreditation surveyors. It is imperative that the selected program area accommodate adequate isolation room if an infection occurs. Administrators should provide an effective set of policies and procedures as well as trained staff to treat infections that may arise. All of these issues must be addressed prior to accepting referrals. Many sources, including the state Medicaid department, can provide information on policies and practices.

REIMBURSEMENT

Identifying potential payor sources is perhaps the single most important determinant of success or failure. Respiratory care patients cost significant dollars to care for. The location of the setting will play an important role in determining reimbursement. Needless to say, a successful program cannot afford to be denied payment for services at any level. A well-rounded approach is often the best plan. It is important to interview the various payors in your area. Find out what their needs are both medically and quantitatively. Typical reimbursement sources fall into the following categories:

- Medicare
- Medicaid
- Managed care HMO/PPO
- Indemnity
- Facility specific contracting with an integrated system or major medical group

Medicare

Medicare has been one of the primary payors in the subacute setting. However, the way in which costs are reimbursed can vary widely. Medicare reimbursement is based on past and future cost report information. After a Medicare program has been established for three years, the cost base is set. Nursing is the largest routine cost area. Medicare imposes routine cost limits (RCLs). RCLs are based on facility and regional data. Daily charges in the routine cost category that exceed the cap are lost and typically unrecoverable to the facility. Upon establishing a subacute care respiratory/ventilator program, the staffing levels for nursing often increase. This can be a significant increase and push costs well in excess of the RCL.

Medicare has a process by which facilities may become exempt from the RCL by demonstrating that they provide an atypically high level of patient care for extended periods of time within a distinct unit. This situation is far from standardized at this time. The HCFA is working toward clarifying the process. The place to begin is with the Medicare intermediary, supplying it with cost reports and patient care data, demonstrating your level of care, and including a written request for an exception from the RCL. If the intermediary agrees with the assessment, it will forward the data to HCFA. HCFA will review the request during the next several months. Then HCFA may approve the request, ask for more information, or deny the request. If the request is approved, the facility will be reimbursed on a cost-based status with no cap of the routine cost center for the year of the filing. The costs submitted on the annual cost report must still meet the criteria of allowable costs. A facility must file the exception request every year in order to maintain the exemption. The primary problem with this approach is that a facility must financially support an expensive level of care for several months while the exemption is reviewed. This can mean several hundred thousand dollars in outstanding charges.

All new providers in the Medicare program have a three-year exemption from the RCL. This allows a facility to develop a program and establish a cost base and an acuity level without being penalized. During this three-year exemption period, requests can be filed and demonstration of atypical care can be accomplished. Once the three years are up, the facility should have established an atypically high cost base and exemption status. Usually a new Medicare provider is either a new facility, a newly licensed unit, or a facility that has not participated in the Medicare program in the past.

Managed Care/Indemnity

The managed care and insurance dollars are the most sought after. Usually, these payors are eager to learn more about all types of subacute programs—especially programs that care for ventilator patients and can demonstrate positive outcomes. Managed care entities and insurance companies can tell you the cost of a ventilator patient in an acute care setting. Subacute settings can offer high technological care at a substantially lower cost.

Payors then have an alternative for the hard-to-place patient. One goal should be to obtain a preferred provider contract for subacute care services for respiratory/ventilator patients. Again, the facility must know the type of patient that the program can and cannot care for. Determine the costs. Decide what your charges will include and exclude, and be specific. Payors of all types want to see progress. Naturally they feel that lower levels of care should come at a lower cost. Therefore, those facilities with a tiered rate structure and tangible qualifying criteria for different levels of respiratory/ventilator patients have success in contracting. The following partial list can assist you in developing your costs and to structure charges:

- ◆ Room and board—Private/semiprivate/isolation.
- ◆ Nursing—Broken down by labor hours per patient day.
- ◆ Therapies—Respiratory, physical, occupational, speech, and recreational, broken down by units of therapy per day.
- ◆ Physician—Varies depending on the facility and managed care payment structure. Just remember to address the cost if it is to be included.
- ◆ Dietary—Include a weekly visit by a registered dietitian. Also, address tube feeding (TPN), and the related equipment to deliver these components.
- ◆ Medications—Differentiate between standard and nonstandard medications.
- ◆ Equipment charges, such as ventilators, feeding and IV pumps, specialty beds, oxygen equipment, and wound care products.

Persons charged with the operation and promotion must be able to address every one of these issues. Information should be readily available for the lengths of stay and discharge options. A facility must begin discharge planning for each patient upon admission. Know what outcomes can be expected and make them specific, measurable, and available. Those who come prepared with the answers will be most likely to walk away with an agreement. Until payors have a comfort level that a facility can actually provide the services they offer, contracting may be done on a case-by-case basis. Specifically, payors will want to see patients weaned off of the ventilator, trained to manage the various types of equipment they will need, and discharged home.

Collaborative Relationships

Lastly, facility-specific contracting may be a possibility for a program. In many parts of the country, hospitals are taking aggressive roles in developing their own insurance products to market to large employers. In this circumstance, the hospital shares all responsibility to provide the most cost effective and appropriate patient care possible. All types of patient care do not necessarily exist within the walls of the hospital. Opportunities exist for a subacute care provider to contract directly with the hospital to provide a lower cost and diagnosis specific level of care that augments that of the hospital.

REMODELING THE FACILITY

Remodeling a facility for a subacute care ventilator program can often be the point at which owners and corporate offices decide to turn around and go home. Costs of a full-scale remodel for a ventilator subacute care program can easily surpass $350,000. These costs depend on

- State regulations
- The age of the current structure
- The chosen location
- The extent of medical and technological equipment
- Types of subacute patients

The primary system that usually needs to be upgraded for ventilator-dependent patients is electrical. However, before you can map out the amount of electrical upgrades needed, it is essential to first address oxygen, air, and suction systems.

There are two basic systems: either in-wall (piped in) or portable. In-wall systems offer many obvious advantages:

- Overall lower cost of operation
- Better use of patient space
- Convenience and ease of operation for staff

- ◆ More flexibility in the types of equipment that can be utilized
- ◆ Cosmetics

These systems are quite costly and may not be possible to install in all building types. Consider oxygen. Bulk liquid systems are far and away the cheapest to operate and offer the most flexibility for clinical use. The running of the medical-grade, high-compression line is the most costly component of installing a bulk system. If dismantling walls is part of a remodel, it makes sense to install an in-wall system. Or better yet, if a new building or addition is planned, it makes fiscal sense to install a bulk system. Location of the bulk tank can pose a problem. Different states and municipalities require varying degrees of distances from living structures to where the bulk tank is placed. This minimum may be as little as 20 feet or more than 60 feet. If your facility does not have the luxury of available space around its building, bulk liquid may not be an option. Check with local fire inspectors and city building offices for specifications. An average cost of a bulk system can be $50,000 to $125,000; however, there are vast regional differences. It is wise to contact a local bulk liquid oxygen dealer for installation and cost information. In some circumstances, a bulk liquid oxygen dealer may provide the tank itself at little or no cost. The dealer can also direct planners to a certified piping contractor.

If an in-wall system is not an option because of cost or regulations, there are acceptable alternatives. In-room liquid tanks and oxygen concentrators offer oxygenation that can meet the needs of most subacute care patients. Concentrator placements are restricted by the use of room space, limited choices in the types of life support equipment, and their high noise level.

In-room liquid tanks are exactly that—tanks that hold liquid oxygen at the patients' bedside. The oxygen can be titrated into the ventilator or directly connected, depending on the brands used. These units make virtually no noise. Their limitation comes from the amount of room consumed at the patients' bedside and that the tanks are unattractive. The tanks are 3–4.5 feet tall and 1–2.5 feet in diameter. Their most significant drawback is that they must be refilled. Depending on the oxygen needs of the patient and the size of tank, they may need to be refilled every three days. This can create a major logistical nightmare as well as a draw on staff productivity. Additionally, local fire codes may prevent in-room location of a liquid tank because they release minute amounts of oxygen into the air.

Concentrators offer the most reasonable alternative to wall oxygen. Good machines can produce five liters per minute of oxygen that is 90%–98% pure. The strongest advantage of concentrators is that they do not run out of oxygen. Their primary drawback is noise. These units use a compressor to force room air through a series of sieve beds that filter out all components of room air except oxygen. The compressors make noise and generate a small amount of heat. During the 1990s, manufacturers have greatly improved the quality of their machines and reduced noise levels.

A good place to begin researching these options is with a local durable medical equipment dealer. Such a dealer can also work out low-cost rental/lease arrangements or direct you to manufacturers in order to purchase equipment.

In-wall suction and air must also be evaluated. Again, the primary issue is cost. Each of these systems must have high-pressure lines run behind the wall similar to that of bulk oxygen. The air system uses a medical-grade air compressor located somewhere within the facility. The wall suction must have a medical grade negative pressure generator within the facility. Each of these units can take up a fair amount of space. Again, these systems offer the most in long-range, lower cost operation and in staff convenience. There are also portable alternatives to these air systems if your facility cannot install them.

Portable compressors and suction machines like concentrators have improved dramatically over the last few years. Certain brands of suction machines can reach the same level of negative pressure as a typical in-wall system. They are very simplistic machines that run for years with low maintenance costs. Portable compressors offer similar quality with little repair costs. However, each piece makes noise when in operation and takes up room in the patient space.

Once a choice is made about which oxygen, air, and suction systems are right for the facility, the question of electrical system capabilities and costs can be addressed.

In addition to the life support equipment, there must be voltage for equipment such as electric beds, room heating and cooling units, and 24-hour emergency backup power. The large voltage requirements of individual heating and cooling units are enough to overload the typically wired long-term care facility patient room. It is easy to see that the electrical component of a remodel is usually the most costly.

Exhibit 3.1 lists the basic equipment and amount of amperes required for each patient space.

Each patient space should have eight available outlets, four of which should be on 24-hour emergency power.

The generator is the most vital item of a life support capable unit. A generator must be sufficient to carry all equipment for the anticipated maximum patient capacity and for an extended period of time. An average unit of 12–15 patients warrants a 450–550 kilowatt generator. These are sizable machines that have a large price tag. Costs for the machine and installation can be $50,000 or more. This equipment also requires space. Consult an electrical engineer as a start. The engineer can provide a detailed list of the equipment and map your current electrical capabilities as well as any future needs. Electrical engineers

EXHIBIT 3.1
NECESSARY EQUIPMENT AND POWER FOR A RESPIRATORY CARE UNIT

Equipment	Amperes
Ventilator	5–7
Oxygen Concentrator	5*
Portable Compressor	2*
Portable Suction	2*
Tube Feeding Pump	2
IV Pump	2
Electrical Bed	6
Total	25 amps/bed

* Items that may not be necessary if using in-wall systems.

can provide information regarding placement of a generator and the fuel source that would be best suited to the specific situation.

Areas such as nurses' stations, shower/bath facilities, therapy rooms, hallways, decor, patient rooms, common space, and lighting are important to the form and function of a respiratory/ventilator subacute care program. It is important for nursing and other staff to be in close proximity to the patients and be able to respond quickly. A patient call system that is tied directly into ventilator alarms is a good idea. Medication and medical supply rooms are also important. Wall units in patient rooms require evaluation, especially if portable types of equipment are going to be used. Units shown in Exhibit 3.2 can either be built into the wall itself or put on wheels to be a part of the room furniture. Designs should keep portable equipment in a neat compartment out of the flow of traffic while still being accessible and usable by the staff.

Attention needs to be given to flooring for a subacute care ventilator program—it's the eternal struggle between carpet and tile. Infection control is a continuous battle in a ventilator program, and carpeting increases the risks of harboring bacteria. Sputum, formula, food, and bodily fluids lead very productive lives in the fibers of carpeting. Despite hours of housekeeping labor on deep cleaning, and inconveniencing both patient and staff, a bacterial analysis can show some unsettling results. There is a growing preference for vinyl tile in the patient rooms and carpet in the hallways and common spaces.

EXHIBIT 3.2

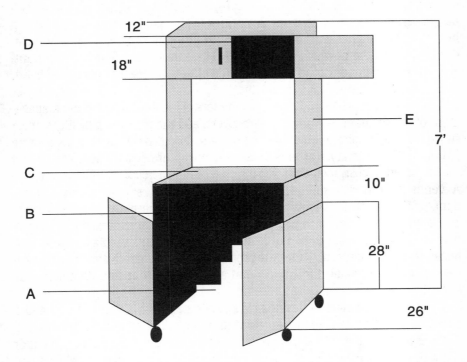

This cabinet is designed to house the following types of equipment: concentrator, ventilator, suction, compressor, and medical supplies. Area A is for the concentrator. It has cabinet doors on the front of the opening to enclose the device. The back and bottom have openings to allow airflow to the concentrator to prevent overheating. Area B is a shelving area for the suction and compressor machines. A 2-inch hole cut in the back panel of this enclosed shelf enables electrical cords to pass. Area C is for the transport ventilator and humidification systems. Area D is a cabinet area with an inside shelf for medical supplies, dressings, and humidification water. The entire unit is on casters to allow easy moving for cleaning. The side panels (E) must be sturdy enough to support humidification brackets and water to be mounted on the side of the cabinet.

STAFF

The introduction of a significantly higher level of care to an established facility often causes a natural stratification among the staff. Managerial personnel can harbor serious reservations about this degree of change. In addition, by bringing in a perceived higher level of staff for a subacute care program, a division can occur between the new and existing staff. These problems are not insurmountable but must be understood and monitored closely before they get out of hand, causing serious management problems.

In staffing a program, first set parameters of responsibility for the management of the program, the primary care givers, and the remaining facility staff. A subacute care respiratory/ventilator program should be distinct in its staffing structure. The primary reason is to accurately capture costs. There are several items related to cost, including nursing staff, that can become difficult to track. A ventilator program should have a dedicated staff, supply room, and patient/staff areas for this reason. Clinical ladders can be built on skills testing. These offer advancement in pay and opportunity to persons who are willing to take on the added responsibilities of higher acuity patients such as those whom respiratory/ventilator subacute care programs attract. This advancement potential can often outweigh the anxiety and ambivalence that staff members may experience when such a program begins.

The place to start in developing a staffing pattern is with a focus and a philosophy. Subacute care is a different kind of patient care environment and requires a somewhat different approach to care delivery. Traditionally, in either acute or long-term care, each licensed professional or department delivers one care component only. Nurses do nursing, respiratory staff do respiratory, and so on. An interdisciplinary approach works best in subacute care ventilator programs. There is one goal for all staff members: to care for and rehabilitate the patient. Everyone shares responsibilities except for those tasks limited to specialists by licensure or professional scope of practice laws. All employees within the program should be able to address issues with reasonable accuracy about any patient. This approach seems to bring out the most staff involvement in patient care.

Twenty-four-hour registered nursing (RN) coverage is standard. This is a component of any true respiratory/ventilator subacute care program. The training of these RNs is primarily acute care oriented with specialized pulmonary, critical, or intensive care training and experience. A comprehensive orientation to long-term care rules and regulations is necessary if the respiratory/ventilator subacute care program is located in a skilled nursing facility (SNF). These persons in essence become the foundation of the clinical interdisciplinary team, and a good working knowledge of SNF patient care guidelines is important. These rules are different from acute care, and ignorance can decrease the chance of becoming successful.

The remainder of the nursing staff can be a mix of RNs, LVN/LPNs, and nurse's aides. Nurse's aides can play an important role in subacute care programs. There is a great deal of hands-on care that must be delivered to patients who are on life support systems. One thing these patients have been lacking in their hospitalizations prior to discharge from the acute care setting is human touch. Someone who takes the time at the bedside to talk with the patient can improve his or her attitudes and speed recovery. Often nurse's aides are among the most eager to meet the challenges that these patients pose.

Another key area in the staffing of a respiratory/ventilator care program is the respiratory therapist. Respiratory therapists have not been common staff members in traditional long-term care settings for two simple reasons: (1) costs of employing a respiratory therapist and (2) Medicare. The Medicare rules that govern cost reimbursement to the facility are complicated and were designed when respiratory therapy was much less a profession than it is today. If a facility were to hire respiratory therapists, the associated costs would have to be allocated under the routine cost center on a cost report. Unlike physical, occupational, and speech therapists, respiratory therapists employed by the facility are not considered ancillary. Naturally this poses a significant problem related to the routine cost limits Medicare imposes. A subacute program requires increased levels of nursing, which is also the largest item of the routine cost center. By adding respiratory therapy to the routine cost center, the facility can easily risk significant loss.

There is another method by which a program can obtain respiratory therapy services. The same Medicare guidelines that place respiratory therapy costs in the routine cost area allow the respiratory therapy costs to be charged as ancillary if the therapists meet the following criteria:

- There must be a contractual agreement between the long-term care facility and an area acute care hospital
- This agreement must allow for the provision of respiratory therapists who are employees of the hospital to deliver therapy in the long-term care facility
- All therapy must be delivered consistent with Medicare's rules and regulations for long-term care facilities
- The therapy must be billed to the long-term care facility under salary equivalencies by state and for the fiscal year

If these requirements are met, the labor and supply charges associated with respiratory therapy for Medicare Part A patients become billable as an ancillary.

This system is limited to the Medicare A population only and will not address all of the costs. Yet, it will cover the majority of costs and allow the facility to take full advantage of the routine cost limit.

Respiratory therapy naturally becomes a vital part of the staff. These persons work closely with nursing and can provide hands-on patient care. Several other disciplines must be involved: Physical, occupational, speech, and recreational therapy all provide a crucial component to any program.

After a staffing plan is developed, training and orientation is the next step. Persons who will work in this type of facility must have a multidisciplinary orientation. It is helpful if all employees have been oriented to each position within the program. They should be familiar with the acute care aspects of patient care and equipment, and with long-term care rules and regulations. The staff should be familiar with the types of equipment they will come into contact with and how it is operated. Another goal is to help staff adjust to a high-acuity-level patient.

After a period of observation, each employee should go through an orientation and skills enhancement training session with the respiratory therapists. The goal here is an in-depth review of ventilators, oxygen systems, respiratory equipment, and pathophysiology as well as emergency procedures. It is important that all members working in the program know the various alarm sounds, the meaning of each one, and the correct action to take. This applies to the DON as well as all staff members, such as housekeeping personnel.

After training and orientation has been completed, each employee should be tested for competency. This test is primarily designed to make sure that employees retain important knowledge of patient care and can apply the important principles in various situations. Competency tests should be given to each employee at least annually. Exhibit 3.3 contains a sample skills test for the care needs of respiratory/ventilator patients.

VENTILATORS

As with most types of medical equipment, the ventilator has progressed in its technology to most closely match different respiratory patterns and needs. The ventilator has definitely come a long way from the days of the iron lung. Ventilators are classified in two ways. The first is by the type of pressure—positive or negative—used to bring air into a person's lung. An iron lung is an example of a negative-pressure ventilator. It artificially creates negative pressure on the chest wall by exerting suction within the tank. The patient's head is outside the tank. Air naturally flows from an area of high pressure to low pressure and thereby causes an air exchange in the patient.

EXHIBIT 3.3
SAMPLE COMPETENCY TEST FOR RESPIRATORY CARE STAFF

Skills Test

Employee Name _____ Date _____

Position_____ Score _____

Circle the letter of the correct answer, fill in the blank, or demonstrate appropriate techniques.

1. Identify the two primary alarms found on a positive pressure ventilator.
   a. High pressure
   b. Medium pressure
   c. Low pressure

2. Can ventilators operate without being plugged into an electrical outlet?
   a. Yes   b. No

3. Demonstrate the proper technique in using a manual resuscitator to breathe for a patient.
   a. Correct   b. Incorrect

4. Demonstrate the correct procedure for identifying and correcting an alarm situation on a patient.
   a. Correct   b. Incorrect

5. Demonstrate the correct procedure for caring for a tracheostomy appliance.
   a. Correct   b. Incorrect

6. Demonstrate the correct procedure for suctioning a patient orally or using a tracheostomy appliance.
   a. Correct   b. Incorrect

7. What are two clinical indicators that a patient is not being ventilated/oxygenated properly?
   1._____   2._____

8. What is the primary risk in positive pressure ventilation of a patient?
   a. Disconnect from life support
   b. Decreased oxygenation
   c. Infection of upper airway
   d. Pneumothorax

9. What are the primary clinical signs of a pneumothorax?
   1._____   2._____

10. Demonstrate the correct procedures to address pneumothorax.
    1._____   2._____

Negative-pressure ventilators remain in use today throughout the United States. They are commonly known as chest shell ventilators and are primarily designed for patients with neuromuscular disease. Negative-pressure ventilators are not as sophisticated in their abilities to meet patient respiratory needs as are positive-pressure ventilators.

Positive-pressure ventilators are the most widely used type of ventilator in acute, subacute, and long-term care. These machines create a pressure greater than atmospheric pressure inside a chamber. These ventilators precisely control the amount of oxygen inside the chambers from the same level of that found in room air to 100% oxygen. Once a specific level of pressure and volume is achieved, a valve opens and gas flows from the positive pressure chamber to the ambient pressure inside the patient's lungs. Positive-pressure ventilators are often controlled by microprocessors and can be minutely adjusted to meet virtually any patient situation.

Within the class of positive-pressure ventilators there are also some variations. The type of ventilator most commonly used in acute care hospitals are the large, multifunction machines. Examples of these types of ventilators are the Puritain Bennett 7200 and the Servo 600C. There is also a class known as transport or portable ventilators. These ventilators are one-sixth the size of the larger ventilators and can perform most of the same functions. Examples of these machines are the Lifecare PLV 102 and the Bear Medical 33. Portable ventilators can also be used with internal batteries for mobile patients, and this feature offers unique advantages. The type of ventilator chosen for use is in part determined by the types of patients admitted and the arrangement and design of patient rooms. Transport type ventilators are able to meet the majority of patient needs in a subacute care setting. However, it is advisable to have access to the more sophisticated machines on short notice.

Ventilators are expensive by anyone's standards. A large acute care-style machine can cost between $15,000 and $30,000. A transport ventilator cost ranges between $5,000 and $9,000. These are alternatives to outright purchases, which are rare. Purchase is typically because of the expense. Rental and lease options are usually the most common arrangements because of the lower capital outlay and maintenance agreements. In either case, be sure to closely review the repair contracts in purchase or lease arrangements.

Ventilators must be serviced at set hourly intervals, and annual maintenance checks can be costly. Maintenance on rental and leased machines should be part of the monthly costs and terms of an agreement. After a machine is owned, a maintenance contract should be extended annually.

Other pieces of equipment are those needed to deliver gases to the patient for therapy, primarily medical air and oxygen. This chapter discussed earlier the pluses and minuses of in-wall versus portable types of equipment. In either case, other items are needed to deliver gases to the patients. Portable compressors, suction machines and concentrators are self-contained and do not require any further pieces of equipment for their use. In-wall systems need metering equipment to deliver the gas from the wall outlet to the patient. Air, oxygen, and suction systems all require an appliance to insert into the wall outlet to adjust their flow and pressure. Equipment such as blood pressure manometers, tube feeding pumps, intravenous (IV) pumps, reclining geriatric chairs, and electrical beds must be available.

SUMMARY

Respiratory/ventilator subacute care programs offer myriad challenges and opportunities. The complexities and costs require careful consideration, evaluation, and capital. With appropriate planning and assessment these programs can benefit a community's needs *and* a corporation's bottom line.

CHAPTER 4

A Wound Care Model

"Captain, it's so obvious we almost missed it!
It's just taking different forms to survive."
— Dr. "Bones" McCoy

Damage to skin integrity in the presence of other factors can result in devastating effects on a patient. The impact of the chronic wound impinges on all human domains. It impacts the physiology by placing a heavy toll on the body's immune system and other reserves in a futile endeavor to heal, and it impacts financially because wounds require costly products, equipment, and labor. Statistics related to wounds are often questionable because many wounds are not reported; however, it is estimated that there are more than 5,000,000 chronic wounds in the United States. Although only a small percentage of these wounds are treated on an inpatient basis, the cost of healing a wound can be high. Healthcare payors express a desire to identify these individuals because they know that the elderly are more likely to develop chronic wounds and that persons with specific conditions, such as diabetes or vascular insufficiency, have a propensity for cultivating chronic lesions. To lend clarity to this task, the *acute wound* is considered a tissue defect that benefits from a simple dressing but would probably heal without one. The *chronic wound*, on the other hand, is the wound that fails to demonstrate progressive improvement within seven days of occurrence.

Using this uncomplicated definition, a profile of persons prone to develop persistent soft tissue lesions was developed. It was determined that as well as people with diabetes and vascular disease, other persons were likely to acquire a chronic wound; people with severe nutritional deficits, paralysis, mobility limitations, or cognitive disorders, as well as the elderly and the immunologically compromised patient. Even routine surgery may result in a chronic wound when a surgical incision, which normally heals steadily and with minimal attention, instead mutates into dehiscence, suppuration, or evisceration.

Since a large number of people hospitalized are funded by Medicare and Medicare extends prospective payment for those patients based on the reason for admission, treatment of a chronic wound such as a pressure ulcer is usually not reimbursed. This is because the chronic wound is not considered relevant to the reason for a patient's admission. This type of funding leaves a hospital liable for the high cost of wound treatment with no opportunity to recover these costs. With a better understanding of the scope of this problem, it was believed that there was a market for a wound care program

- Provided it could standardize wound care delivery
- In a setting where funding could be more effectively managed
- Where wound healing and recovery of the individual with a wound could be conservatively priced

Experience with nursing homes had established that a minimum of 10% of residents developed pressure ulcers. This offered additional evidence of a healthcare setting that could benefit from a specialized wound treatment center. The wound care program could seek referrals from hospitals and nursing homes.

THE MODEL

A model was designed to comprehensively address the needs of the patient with a chronic wound. The lower overhead of the nursing home was appealing from a cost implementation perspective. An abbreviated length of stay was desirable, and the goal-oriented, multidisciplinary team approach was recognized as essential. It was observed that patients in long-term care subacute settings were faring as well as patients in the hospital, for substantially less money.

This prompted some healthcare planners to begin designing, developing, and implementing a subacute wound program in nursing homes. Using a certified enterostomal therapist (an RN with advanced training in wound, ostomy, and continence management) to assist in defining and designing the prototype provides the program with guidance from an expert wound care resource. This professional is adept at appraising and using the myriad of wound care products available and is knowledgeable in the most effective, lowest cost interventions to promote tissue repair.

One objective for the program was to avoid a focus on pathology and instead approach the patient as a unique and total being, a participant in a community of family and friends. This approach mandated a program to address not only the wound, but all influences on the patient:

- Self-care abilities
- Nutrition
- Support systems
- Home setting
- Mobility
- Economics
- Coping skills
- Understanding of wellness

THE PROGRAM

The mission of a subacute wound care program is to provide comprehensive, scientifically based care for the recovery, restoration, and rehabilitation of the individual with a chronic wound in the most cost effective manner while striving for maximum patient and family satisfaction.

Elements of Program Development

Many factors contribute to success; however, clinical and financial performance are most closely linked to the following:

- A comprehensive staff orientation to all aspects of Medicare reimbursement
- A complete analysis of the cost of necessary supplies, products, services, procedures, and employment of a supply person competent to establish and maintain sophisticated tracking systems
- Twenty-four-hour staff with IV certification
- Full time, facility based therapists with rehabilitation experience
- Business office staff skilled in complex insurance and Medicare billing
- Board-certified enterostomal therapy nurse (CETN)
- Medical director evaluating patients within 24 hours of admission and rounding at least twice weekly
- Case manager
- Marketer
- Social worker
- Orientation for staff, residents, families, providers, and community
- Customer service training for staff
- Clerical support

Scope of Care

Subacute wound care programs provide the following specialized assessment and interventions to the individual admitted:

- ◆ Management of complications such as failure to heal, dehiscence, abscess, evisceration, undermining, tunneling, suppuration, infection, fistula, or other pathology
- ◆ Palliative and/or compassionate management of pervasive tissue destruction to contain drainage and odor, protect viable surrounding tissue, control pain, facilitate homeostasis, and eradicate infection
- ◆ Management of lower-extremity ulcers associated with vascular or neuropathic disease
- ◆ Management of defects resulting from chemotherapeutics
- ◆ Management of tissue destruction associated with osteomyelitis, gangrene, malignancy, ecthyma, erysipelas, herpes, cellulitis, foreign body, or allergic response
- ◆ Management of skin grafts, muscle flaps, or tissue expanders
- ◆ Management of damage from drainage at stomal or percutaneous tube site or other severe pathology
- ◆ Management of drains, tubes, wounds, pouching systems, or other specialty needs for heavily exudating wounds
- ◆ Management of complications related to amputation, surgery, radiation, trauma, prosthetics, orthotics, etc.
- ◆ Management of thermal or chemical injury

Expected outcomes from treatment in a subacute wound care program are improved quality of life through participation in ongoing, consistent health maintenance, and improved skin integrity. Patients should recognize the role of prevention and positive personal health choices in achieving optimum wellness. They should have enhanced knowledge of self-care, skin inspection, and recognition of signs of impending loss of tissue integrity as demonstrated by successful completion of a training program. Patients should understand the role of nutrition, hydration, and elimination in maintenance of skin integrity. They also need to acknowledge and recognize necessary lifestyle changes, including those related to role adaptation, support mechanisms, body image, and coping abilities.

Admissions Standards

The individual admitted may demonstrate an integumentary defect requiring intensive physical rehabilitation, aggressive nutritional therapy, specialized nursing intervention, extensive interdisciplinary team care, and comprehensive medical management. This can include defects such as a full-thickness lesion (stage III–IV) or multiple, shallow stage I lesions located such that patient cannot be positioned off these defects. The defect may be a result of trauma, or acute or chronic illness.

Transfer and Discharge Criteria

Transfer and discharge of the patient will occur when the goals established are achieved, the patient's condition alters requiring hospitalization, or the patient refuses to participate in the agreed upon plan of care.

Administrative Standards

Before implementing the subacute wound care program, administrative standards must be established. These standards identify the program components and document the unit's commitment to a level and quality of care. A manual including policies and procedures should be available to all program employees.

Administrative standards minimally address

+ The program's mission and philosophy
+ Patient outcomes
+ Admission, transfer, and discharge criteria
+ Reimbursement
+ Billing
+ Staffing and clinical practice
+ Scope of care
+ Team function and purpose
+ Infection control
+ Quality
+ Patient/family education
+ Staff development

Staff Specialists

Key constituents of successful development and implementation of the subacute wound care program are the experienced program developer and the experienced board certified enterostomal therapy nurse (CETN). The CETN is critical to the success of the program because this practitioner, by training and education, is able to provide not only advanced wound care, incontinence recommendations, and ostomy management, but also data by which financial and clinical progress is measured. This clinician has specialized training and knowledge in the appropriate use of wound care products and procedures and the measures necessary for obtaining reimbursement for them. With recent changes in the *Medicare Carriers Manual*, the CETN is pivotal because this practitioner, depending on state requirements, may perform debridement as needed. Such treatment affects reimbursement.

The CETN should meet the following performance expectations:

1. Determines and reports pressure ulcer incidence and prevalence within the facility before the program opens in order to establish a statistical baseline allowing regular comparison to statistics obtained after opening. This information documents improved patient conditions and provides a method of noting outcomes, trends, or problems and addressing them promptly. These data are extremely valuable in marketing to payors and other referral sources.

2. Evaluates and educates personnel on the efficacy of wound care products and equipment. This facilitates cost-efficient use of supplies and promotes team accountability for expenses.

3. Eliminates redundant, unnecessary, or excessively expensive supplies, thus lowering overhead.

4. Creates (with the central supply representative) an organized and appropriate inventory of wound care products and equipment; and assists the central supply staff in implementing a data collection system to evaluate usage, trends, problems, and so on. Appropriate inventory lowers overhead, prevents products from expiring on the shelf, and allows efficient retrieval and tracking.

5. Determines (with appropriate team members) suitable enteral formulas, products, and delivery systems (pumps, tubing, and so on); reviews and updates policies and procedures for enteral delivery; reviews billing procedures; provides education; develops programs for business, nursing, and central supply personnel; and assists the central supply staff in developing a master list with pricing and other pertinent data for reference and ordering.

6. Works with the education department to ensure staff understanding of and compliance with medical requirements and ensures that appropriate documentation is achieved.

7. Works with central supply staff, the director of nursing, clinical manager, and administrator to review vendor contracts, establish new contracts as needed, and promote a close working relationship between vendors and administration.

8. Arranges for staff inservice programs about products and equipment by vendors before the program opens, and on a regular basis thereafter.

9. Assists the central supply representative in development of a master list of vendors, supplies, and equipment and other critical purchasing or leasing information to provide an organized and systematic method to obtain products. This allows central supply to operate when the central supply representative is out of the building. Because of their product and service orientation, enterostomal therapists (ETs) typically have exceptional vendor relationships. ETs can often obtain better prices on products, including specialty beds. This can provide substantial savings to the facility.

10. Reviews wound care policies and procedures to establish that they are in accordance with safe and accepted practice guidelines; establishes standards for managing contaminated and infectious body fluids and waste; and oversees disinfection of wound care equipment utilized by the team.

11. Establishes inservice education to address infection control issues; and communicates and works with the infection control personnel in these matters. This procedure ensures that clinical standards meet the national and community norm and are founded in science and research; that they protect the staff and patients from contamination by infectious agents; and that they reflect reimbursement standards and cost effectiveness in the directions for their use.

12. Evaluates therapist's and nursing staff's competence in wound care delivery, rectifying deficiencies, and annually repeats and documents these evaluations. Documents that care givers have been trained, certified, and reassessed according to printed standards established as clinically sound and appropriate.

13. Using a surgical instructor, establishes training regarding debridement for licensed personnel (as eligible by state practice). Sets standard for yearly reevaluation of all such practitioners. Provides documentation that wound care practices are based on accepted standards.

14. Performs sharp debridement and other types of debridement as indicated and as dictated by the state practice act. This saves the high cost of surgeon's debridement and facilitates wound healing and reimbursement when surgical dressings are indicated.

15. Reviews all standards of wound care practice and procedures, updating, revising, and assisting the educational department in educating all staff. Standardizes the wound care practice within the facility to ensure safe and appropriate clinical practice.

16. Arranges educational programs for state and/or federal surveyors on policies and procedures, and the purpose to which the program adheres. Surveyors are less likely to find fault if the goals and objectives of the program and its associated activities are explained.

17. Establishes a working relationship with the fiscal intermediary and carrier to enhance communication and appropriate use and reimbursement for products and services. The Wound Ostomy and Continence Nurse's Society has shaped wound care reimbursement standards through education and open communication with the Health Care Financing Administration (HCFA) on a national level, obtaining more appropriate coverage for patients. The individual CETN extends those efforts on a local level to benefit the patients and others in the community.

18. Evaluates wounds for infectious process. Overzealous or inappropriate culturing of inflamed wounds is an unnecessary expense to the patient and the facility. The results of these inappropriate cultures often lead to antibiotic regimens that are ineffective, expensive, and further debilitating to the patient. The CETN educates staff on the differences between infection and inflammation and the appropriate interventions.

19. Evaluates bowel and bladder management programs, and assists in implementing continence rehabilitation programs. Effective bowel and bladder management can save a facility thousands of dollars each month and may allow patients to be returned to an active role in society. The CETN has advanced knowledge of methods to improve continence.

20. Assists in creating and enhancing the marketing plan; participates in marketing activities as indicated. The CETN and other specialty staff members can be credible marketers.

21. Coordinates regional wound care seminars for providers and professionals. Seminars are an excellent tool to increase visibility of the facility, encourage area referral sources and professionals to utilize the facility, recruit staff, and bring in revenue.

These capacities, in which the CETN is well suited to function, make this role crucial to the clinical and financial success of the subacute wound care program.

Staff Education and Orientation to Rehabilitation and Wound Care

Operating from a shared knowledge base is fundamental to the proficient functioning of the team. Besides the basic orientation required in any healthcare setting (safety, confidentiality, infection control, documentation, attendance, patient and resident's rights, and so on), there is a substantial need to raise the performance level of the team to one capable of delivering an exceptional product, in an alternate environment, more efficiently, at a lower cost, and with greater customer satisfaction. To meet these objectives, orientation and training courses are suggested for the staff such as those listed in Exhibit 4.1. Along with education guides, lists of references, books, and journals are necessary for the staff to deliver the best possible and most advanced care. A suggested reading list for wound care appears at the end of this chapter. It is important to hold frequent seminars and educational programs for the personnel. Exhibit 4.2

EXHIBIT 4.1
SAMPLE EDUCATIONAL PROGRAM FOR FUNDAMENTALS OF WOUND CARE

Fundamentals of Wound Care

Intended audience: Members of the interdisciplinary team

Objective: At completion of this program, the participant will score 80% or better on a comprehensive test of the information.

Unit I.	Psychology of wound healing
Unit II.	Chronic wounds: causes and risk factors
Unit III.	Nutrition and wound healing
Unit IV.	Wound care products
Unit V.	Diabetic and vascular ulcers
Unit VI.	Management of percutaneous tubes (PEG tubes, chest tubes, jejunostomy, and so on)
Unit VII.	Pressure reduction and pressure relief
Unit VIII.	Standards of practice and policies and procedures
Unit IX.	Patient education and discharge training
	Post test
	Awarding of certificates

---

contains samples of subacute wound care standards that could provide education topics.

EXHIBIT 4.2

CLASSIFICATION TOOL FOR STANDARDS OF SKIN AND WOUND ASSESSMENT

STANDARDS OF SKIN AND WOUND ASSESSMENT

This classification tool is applicable to all wounds, regardless of etiology

Stage I.	Nonblanchable redness of unbroken skin. This stage can indicate the first sign of superficial damage reversible with prompt intervention, or it may present the first indicator of deep and pervasive tissue destruction that is irreversible. This stage presents as intact skin that is warm, red, inflamed, tender, and swollen.
Stage II.	Partial-thickness loss of skin involving epidermis. May involve damage into but not through dermis. This stage resembles a blister or an abrasion, is painful, red, swollen, warm, and has minimal drainage.
Stage III.	Full-thickness loss of tissue with damage to subcutaneous structures that may extend down to but not through fascia. This stage presents as a crater and may include undermining, drainage, and necrotic tissue. The wound base is not typically painful.
Stage IV.	Full-thickness tissue loss with deep destruction, tissue necrosis, involvement of muscle, bone, joint, or tendons. Presents as deep crater. Usually has large amounts of drainage. May be covered with necrotic tissue or involve sinus tracts and fistulas. Wound is usually not painful.

Developed by the Wound, Ostomy and Continence Nurse's Society (1986) and adopted by the National Pressure Ulcer Advisory Panel (1989).

THE PHYSICAL PLANT

The primary changes necessary to the facility in which the program will be located are in the areas of administrative offices; therapy space, including equipment and supply space; nursing station; equipment and supply storage; entrance; dining area; and family sitting area. Any of these alterations may require cosmetic or structural modification, reallocation of existing space, or construction of new space. To address physical plant changes, an architect and project manager with healthcare building expertise, and knowledge of federal and state regulations as well as building and safety codes are valuable resources. There is always reluctance to sacrifice the revenue acquired from a patient room to procure additional space. Profits produced by full occupancy of the room at the private pay rate compared with projected therapy revenues from that space can guide the leadership in justifying reallocation of space. An experienced subacute program developer can accurately define these parameters. The following guidelines are based on a 15–20 bed subacute wound care program.

Therapy Space

The gym or other area should accommodate a wet treatment area where hydrotherapy, wound irrigation, debridement, wound dressings, splinting, and casting can be done. A section holding a small refrigerator/freezer for cold treatments, hydroculator unit, disinfection and sterilization supplies and equipment, portable tank, portable suction, and other electronic equipment are also necessary in this area. At least two private patient treatment areas are suggested. To control odor from these treatment areas, the installation of an exhaust fan is suggested.

Locating additional storage space for dressings, instruments, wound care products, linen, disinfection supplies, and protective paraphernalia eliminates the loss of valuable, reimbursable therapy time spent retrieving supplies from the nursing unit or central supply room. Access also allows for more efficient tracking of charges and materials under separate accounts, and decreases procedure preparation time. This allows for more time to be spent with patients. The wet area described here does not replace the gym area allotted for individual or group therapy activities. Space must also be available for groups involving five or more occupied wheelchairs.

Patient Care Area

Once the location of the 15–20 bed program area is designated, modifications are needed to distinguish it from the rest of the center. The modifications can be substantially less complicated and cost less than that for a ventilator program. The following alterations are suggested:

- Four double-receptacle, emergency generator-connected, ground fault-interrupted plugs at the head of each bed
- An emergency generator to supply electrical needs
- Four electrical outlets opposite each bed for other equipment use
- Over-bed exam lighting and glove dispenser at each bed
- Sharps disposal container in each room
- Small locked cabinet at bedside for wound supplies
- Small locked cabinet at bedside for patient's use

Handwashing areas in each room and throughout the facility must be especially well planned. The paper towel container should not require any hand contact to dispense towels, and the dispenser should be positioned in such a way that removing or disposing of towels does not expose hands to contamination from items such as a towel bar or soap dispenser. Trash cans should not require hand operation, again, to prevent skin contamination during disposal. Call lights, telephone, and intercom system require evaluation to determine their adaptability to the new program.

Quality window treatments, wallpaper, privacy curtains, and bed spreads are needed. Electric beds and a comfortable recliner at bedside for family use are also assets. Changing the appearance of the patient room doors as well as nameplates, room numbers, and other signage will help to set the unit apart as a separate entity. A patient/family waiting room must be provided and if possible an additional dining and social area created for the subacute patients. The square footage is dictated by the size of the patient population and any regulations.

A nursing station with seating and documentation space is necessary. This station should have a padded floor to reduce noise and be positioned so all call lights are visible.

The Entry

Following are suggested design elements for the entry to the unit:

- The entry and the approach to the entry should be smoke free
- The colors, furnishings, plants, and accessories of the entry are indicative of a possibly younger and more diverse patient population
- The entry does not serve as a waiting or sleeping area, or a gathering place for agitated residents
- There is signage directing the visitor to the different areas of the center.
- There is a receptionist to assist and screen visitors.
- Parking is convenient and wheelchair accessible
- Walking from the entry to the subacute wound care program space requires minimal exposure to other patient care areas

Cosmetic Improvements

To evaluate the marketability of the physical plant, facility leadership is advised to tour other centers in the area. Observations made with such tours usually establish the need for cosmetic improvements to broaden the appeal of the facility. Since competition is no longer confined to nursing homes, improvements must encompass the facility's exterior appearance, parking areas, landscaping, signage, and the very name of the facility. It is difficult to present as a provider of short-term rehabilitation if the facility name includes terms such as *rest, final, asylum, home, sanctuary, retirement, retreat, eternal, haven,* or *peace.*

Colors, artwork, and decor should reflect rehabilitation, recovery, and community involvement by the center and its patients. The facility must clearly seek to reflect a diverse patient population and the associated interests and activities. In short, the center establishing or redefining itself as a subacute care provider cannot simply rely on its reputation for quality services. It must also have curb appeal to succeed in the dynamic and diverse healthcare market of today. Exhibits 4.3 and 4.4 list supplies and equipment that may be necessary to a wound care program.

MARKETING

Successful marketing depends on the precise and detailed internal analysis of the facility and the need in the marketplace for a wound care program. With this information a strong and dedicated marketing plan will guide the program in identifying its competition, understanding its patient population, and meeting the needs of its referral sources. There are challenges that marketing for a subacute wound care program must squarely meet to succeed. Marketing for subacute care programs is discussed in detail in Chapter 9.

EXHIBIT 4.3
BASIC WOUND CARE SUPPLIES

Permanent, felt-tip markers	Exam gloves (vinyl & latex)	Scissors (disposable & reusable)	Sterilization & disinfection capabilities
Sterile saline	Sterile gloves		Electrodes
Irrigation syringes	Safety pins	Emesis basins	
	Tape (variety)	Disposable razors	Elect. conduct gel
Irrigation kits	Skin sealants	Betadine	
Ziplock baggies	Impermeable	Swabs	Drain sponges
Transparent page protectors (for wound tracing)	protective gowns	Sterile 4 × 4s	Bulk 4 × 4s
	Suture removal kits	Self-adherent compression wrap dressing	Full-face shields (cover forehead to clavicles)
Skin bond cement	Positioning devices	Aluminum foil	Plastic wrap
			Heel elevators

EXHIBIT 4.4
EQUIPMENT AND SUPPLIES LOCATED IN PATIENT CARE AREA

Wet treatment area	Mat table	Portable privacy screen	Mayo table
76 × 48 locked supply cabinet	Privacy screen	Extremity whirlpool (portable, high/low)	Autoclave or other sterilization capability
Exhaust fan	Portable ultrasound		
Clean utility room	Portable E-stim	Full length mirror	Quad mirrors
Dirty utility room	Portable suction	Utility carts	Disinfection supplies
Hopper	TENS unit	Stethoscope	Video camcorder
Sink	Doppler	Treatment carts	IV poles
Hydroculator and packs (varied)	Wound irrigation device carts for electrical equipment	Recliner/lounger	Exam lighting
Wall-mounted TV (with VCR)	Segmental compress, therapy unit	Adjustable treatment stools	Foot-operated trash containers
Watersafe outlets	Sphygmomano-meter leg/arm size	Polaroid wound camera/film	Phones, emergency call system
	Large wall clock	Timer	Sharps disposal unit
		Glove boxes	
		Refrigerator/ freezer (small)	

REIMBURSEMENT

Along with the clinical component, another element critical to success is reimbursement. The three most common funding sources are Medicare, Medicaid, and private or commercial insurance. To maintain profitability, a thorough understanding of the restrictions and limitations of each funding source is essential.

Medicare

A high percentage of persons afflicted with chronic wounds are older than age 65. These individuals have usually been admitted to the hospital for medical reasons that are complicated by the development of pressure ulcers. These frail elderly are often not stable enough for home placement and require additional treatment in a skilled nursing facility. These patients are excellent candidates for the specialized services of the subacute wound care program. For such Medicare patients, reimbursement of covered costs is available for the first 20 days following a three-day hospital stay. With that time frame in mind, team strategies must be carefully planned to maximize patient outcome. If, following a physical assessment of the patient, the team determines that achieving the rehabilitation goals will necessitate more than 20 days, a copayment source must be established prior to admission.

Accurate verification of eligibility, coverage, and resources for coinsurance payment are critical, and for these reasons, the internal case manager and marketer must be extremely proficient. These patients and their families or representatives must be fully apprised of the purpose and the goals of the admission.

In some instances, the patient may not be discharged home for an unforeseen reason. Due to this possibility, alternative discharge avenues must be discussed with and understood by the responsible parties. Success depends on the flexibility and availability of program beds; thus, each person, family member, and staff must comprehend that patients are not treated like residential or long-term patients. These beds and services are available to patients able to participate in an intense program focused on rapid recovery and rehabilitation. Patients unable to be discharged home should be eligible for transfer to a lower level of care setting within or outside the facility, depending on bed availability. Careful research and planning by the case manager prior to patient acceptance will prevent many misunderstandings.

Another crucial element to appropriate admission and discharge planning is uninterrupted, private preadmission and admission process. The case manager should review these issues as well as the policies and routines of the program with the family or responsible individual. Careful attention should be given to reducing stress and anxiety surrounding this meeting, while sharing salient information with the patient and family. This briefing should include what Medicare does or does not cover. It is always advisable to avoid grouping financial issues with other weighty topics such as advanced directives. During this meeting, the patient and family should be provided with verbal and written orientation information.

Medicare patients may comprise a large percentage of subacute care patients, so it is important that those patients that are accepted be appropriately placed. The number of days of Medicare coverage available must be verified, and all details concerning length of stay, discharge and expected outcomes should be thoroughly established prior to admission.

Before the Medicare patient is admitted, the following data should be established:

- Copayment terms
- Daily team intervention to promote maximum independence
- Need for intensive training in: bowel/bladder management, mobility, medication, adaptive equipment use, safety, nutrition, and skin management
- The patient's need for daily therapy
- Patient and family's understanding of admission, length of stay, discharge disposition, and projected outcomes

Medicaid

Medicaid is an assortment of state-administered programs providing assistance to impoverished, disabled, or elderly individuals. Recipients may include those receiving Aid to Families with Dependent Children or those receiving benefits from the Supplemental Social Security. It is an amalgamation of services with varying purposes, administered at the state level, and coverage varies greatly from state to state. For financial viability, an organization considering a subacute wound care program should thoroughly evaluate the state's

- Eligibility requirements
- Reimbursement schedules and rates
- Scope and duration of service coverage

Once it is determined by the facility which services are reimbursable, a cautious analysis should be performed related to the provider's responsibilities under Medicaid requirements. Enhanced coverage for some specialty programs is available, but agreements are negotiated program by program. Healthcare officials are recognizing the subacute care setting as offering specialty services at reduced costs outside the hospital setting. To capitalize on this, the facility must approach the state Medicaid administrator with well-researched evidence of what the subacute care program offers. A presentation that includes the following could result in much improved per diem payments for such patients:

- Statistics related to chronic wounds
- Current costs of treatment utilizing existing resources
- Impact on employment contrasted with projected savings
- Successful rehabilitation of persons with chronic wounds
- Planned return to the work force of the individual

This type of negotiation has proven successful for other subacute care specialties, allowing the provider the resources to deliver a better quality of service.

Prior to a Medicaid admission, intake staff should use a tool like the Medicare preadmission questionnaire to obtain the necessary information. Before the patient is accepted, staff should determine the patient's eligibility as well as duration of coverage. Keep in mind that your best opportunity is to try to obtain a Medicaid per diem for a specialty patient group.

Managed Care/Indemnity

With private insurance or health maintenance organization (HMO) coverage comes another payor type. The individual with chronic wounds who has insurance coverage may be the young to middle-aged adult who has a surgical wound or paraplegia and pressure ulcers. A number of these individuals typically have a need for chronic wound treatment.

These persons may have an insurance company case manager overseeing medical management. Referrals of these patients have been obtained by marketing to these case managers and insurance carriers. A strong incentive to send a patient to a subacute wound care program is a less lengthy admission process that does not require hospitalization and surgical expense. Regular communication with the case manager is essential, and outcome objectives must be met.

In addition to healing the wound, staff assist the patient to regain strength and nutritional equilibrium and provide extensive education and retraining all at a markedly lower cost than an acute care facility would charge. Successful integration of insurance patients requires a knowledge of the cost of each product, service, medication, treatment, procedure, and unit of therapy. Per diem rates and other types of pricing are based on the detailed analysis of every provision that the program extends to the patient. This analysis and a thorough understanding of pricing must be functional long before the first admission occurs. This enables the program to calculate the cost of servicing a specialty bed or a specialty call bell while remaining competitive and profitable. Along with pricing, the program has to make the case managers believe that the subacute care team will deliver a quality service comparable to the acute setting that is satisfying to the patient and family. Demands often heard from insurance case managers include ones for:

- Permanent, full time therapists employed by the program
- One person with whom to communicate on a regular basis
- Consistent staff
- JCAHO/CARF accreditation
- Staffing levels at subacute standards (not nursing home level)

- No surprises
- Delivery of the promised services

Private insurance patients are attractive, but the staff must recognize that although the first referral may not prove difficult, the second referral is impossible if the carrier and the patient are dissatisfied. Of course, there are other funding sources to identify, and they must be approached individually in each community. Preparing the team for the types of patients that will be admitted and instilling in team members a comprehensive knowledge of Medicare, Medicaid, and private insurance reimbursement will significantly assist the program to become fiscally and clinically successful.

SUMMARY

The material presented in this chapter illustrates the investment required—time, money, and human resources—to develop a subacute wound care program. The intent was to also provide information usable for those interested in preparing for a subacute wound care career. Subacute implementation is an absorbing and interesting venture for many. Successful subacute centers thrive with the guidance of the right people. The crux of a subacute care program is its personnel. The care and services extended in this setting are as rigorous as many of those in an acute hospital. Subacute wound care programs offer the provider another source of revenue, the patient a quality healing alternative, and the payor a cost efficient method to provide for its enrollees.

SUGGESTED READING

Journals

- *Advances in Wound Care,* S.N. Publications, Inc. 103 N. Second Street, West Dundee, IL 60118
- *Journal of Wound, Ostomy, and Continence Nursing,* C.V. Mosby Co., 11830 Westline, Industrial Dr., St. Louis, MO 63146
- *Ostomy/Wound Management,* Health Management Publications, 649 S. Henderson Rd., The Westover Bldg., King of Prussia, PA 19406
- *Rehabilitation Nursing,* Association of Rehabilitation Nurses, 5700 Skokie, IL 60077
- *Wounds,* Health Management Publications, 649 S. Henderson Rd., The Westover Bldg., King of Prussia, PA 19406

Books

- *Chronic Wound Care,* Krasner
- *Acute and Chronic Wounds,* Bryant
- *Standards of Wound Care,* Wound, Ostomy, and Continence Nurse's Society
- *Pressure Ulcers,* Maklebust, Sieggren
- *Guidelines for Prevention and Treatment of Pressure Ulcers,* Agency for Health Care Policy and Research
- *Wound Healing: Alternatives in Management,* Kloth, McCulloch, Feedar
- *Rehabilitation Nursing: Concepts and Practice* (second edition), Rehabilitation Nursing Foundation
- *Drugs and Nursing Implications,* Govoni & Hayes
- *Standards of Care, Dermal Wounds: Leg Ulcers,* Wound Ostomy & Continence Nurse's Society
- *Standards of Care, Dermal Wounds: Pressure Sores,* Wound Ostomy & Continence Nurse's Society
- *Pain, Clinical Manual for Nursing Practice,* McCaffery, Beebe

PART III

◆

SUBACUTE CARE CONSIDERATIONS

CHAPTER 5

Financial Implications

"Interns can't accept money."
— Dr. Kildare

Subacute care offers potential financial rewards to providers. As a less expensive level of care, matching these services to the acuity needs of the targeted patient population often produces clinical benefits comparable to those of acute care. In order to truly measure the potential, careful financial evaluation needs to be undertaken. In addition, an in-depth understanding of the financial mechanisms needs to be developed.

THE POTENTIAL

Settings for the provision of subacute care may include freestanding and hospital-based skilled nursing facilities (SNFs), acute care rehabilitation, acute care medical/surgical facilities, long-term hospitals, and specialty hospitals. Many of the proprietary nursing home chains are developing and marketing subacute care services in their facilities. A sense of competition has emerged as both hospitals and nursing homes vie for expedient implementation of a subacute care level of care. There is room for many types of subacute care programs and the license of the parent facility is less important than the services and capabilities of the program.

Historically, hospitals have looked at the development of distinct-part skilled nursing beds as a minor component of their business, largely serving as a relief valve for Diagnosis Related Group (DRG) based prospective payment. In order to meet the needs of subacute care patients, hospitals have increased the amount of ancillary services and nursing care delivered in these programs. In addition to the use of distinct-part skilled nursing beds, many hospitals are utilizing beds that traditionally have been used for general medical/surgical and rehabilitation patients to provide more cost-effective subacute care services to managed care and commercially insured patients.

The growth of subacute care is further evidenced by companies that convert acute care hospitals into long-term hospitals that care for ventilator dependent and other respiratory patients as well as offer subacute care services to augment their other services.

Although conceptually subacute care is a logical mode of care delivery, reimbursement mechanisms for this level of care are not well defined. Managed care and commercial payors are recognizing the financial benefits of subacute care, and providers are experiencing an ever-increasing volume of these patients. An acute care provider with access to a subacute care level would be more attractive to managed care programs by demonstrating the availability of a continuum of care that can lower the total cost of the inpatient stay.

Medicare, which represents a large portion of the current subacute care market, recognizes subacute care higher service levels through beds licensed as freestanding or as a hospital-based skilled nursing facility, as well as through long-term hospital reimbursement mechanisms. All of these developments result in a potential for providers to maximize operational efficiency as well as financial performance.

TYPES OF REIMBURSEMENT

As mentioned previously, reimbursement by Medicare represents much of the subacute care revenues. Under the current reimbursement system, three Health-care Financing Administration (HCFA) designations exist as avenues for pay-ment: freestanding skilled nursing facility, hospital-based skilled nursing unit, and long-term hospital.

Although all three methods of Medicare reimbursement are cost based, rather than prospectively reimbursed (DRG-type payments), they differ in their methods of calculating the reimbursement amount and associated operating incentives. The first two options, freestanding and hospital-based designated skilled nursing care, are reimbursed with predetermined limits based on per diem inpatient routine service costs, with ancillary services and capital costs reimbursed based on reasonable charges.

The long-term hospital designation is applied to a facility that has an agreement with Medicare to participate as a long-term hospital with an average inpatient length of stay greater than 25 days. Reimbursement for this type of facility is similar to rehabilitation hospitals and units whereby a base year cost-per-discharge is determined in a given year, and that amount, also referred to as the TEFRA (Tax Equity and Fiscal Responsibility Act) target amount, becomes the reimbursement basis per discharge in subsequent years. The target amount is adjusted annually by a market-basket factor. The provider receives the lesser of the actual cost per discharge or the target amount. If the provider is able to deliver the care for less than the target amount, the provider receives an additional incentive payment. If the provider's cost per discharge to deliver the care exceeds the target amount, the provider receives the lesser of the ceiling plus half the excess costs or 110% of the ceiling amount.

For the most part, facilities designated as long-term hospitals represent a small amount of subacute care providers. Recent initiatives by HCFA have focused on revising the incentives associated with long-term hospital reimbursement. These incentives include a "hospital within a hospital" concept. To achieve this designation, several requirements must be satisfied, including a separate governance structure. Exhibit 5.1 summarizes all three settings.

EXHIBIT 5.1
SUBACUTE CARE MEDICARE REIMBURSEMENT MECHANISMS

Type of Setting	Type of Licensed Beds	Medicare Program	Payment Method
Freestanding skilled nursing facility (SNF)	Long-term care	Skilled nursing, Medicare certified	Subject to routine cost limits after third year; unlimited ancillary; 100 days coverage with coinsurance after 20 days—may apply for exemption from limits
Hospital-based/ SNF unit	Long-term care	Skilled nursing, Medicare certified	Same as freestanding nursing home, but there is a preadjustment
Long-term hospital	Acute, specialty, or chronic hospital	PPS exempt as long-term hospital with average length of stay > 25 days	Under TEFRA reimbursement—must include total hospital; returns to PPS reimbursement if length of stay drops below 25 days

Comparing and contrasting the two types of Medicare reimbursement reveal the following differences in incentives:

◆ Long-term hospital patients must stay a minimum of 25 days under the Medicare program, whereas SNF patients have a maximum 100-day length of stay and must be admitted from an acute care hospital following a three-day stay.

◆ Under the long-term hospital program, no difference is recognized between ancillary and routine costs, as compared to a SNF, which limits routine costs but not ancillary services as long as they are medically necessary.

◆ Long-term hospitals recognize the total cost of care during the length of stay, whereas SNFs have per diem limitations.

◆ In comparing total reimbursement under the Medicare program on a per diem basis, Medicare revenues would tend to be higher for long-term hospitals than for freestanding or hospital-based SNFs.

◆ These differences in reimbursement tend to favor long-term hospitals in some ways as a treatment alternative for medically complex patients.

In addition to Medicare, Medicaid programs in several states are recognizing the value and cost savings potential of caring for medically complex patients. States such as California, Illinois, Ohio, Nebraska, Maryland, Pennsylvania, Delaware, and Virginia have designated or are creating reimbursement programs for subacute care. For the most part, the reimbursement mechanism in these states is aimed at providers of specialty programs. Criteria related to equipment, staffing, and physical plant requirements are specific to each state's programs.

Managed care and commercial insurers are also contracting for this level of care. This phenomenon has had a dramatic impact on both acute care and rehabilitation case management. Managed care and commercial insurers are structuring reimbursement incentives that foster early discharge from acute care and/or frequently bypass acute or rehabilitation facilities entirely in favor of subacute care programs. In some markets, managed care has completely redirected specific types of medical and rehabilitative cases, creating, in effect, subacute care critical pathways. Typical pathways are characterized by a hip replacement patient staying only three to four days post-surgery before transfer to the less expensive subacute care provider for therapy.

PROJECTING SUBACUTE CARE FINANCIAL PERFORMANCE

In evaluating potential for subacute care development, a realistic projection of revenues, incremental costs, and total costs is required. The analysis should consider staffing, ancillary, and capital costs required for development. An evaluation of both the contribution margin (variable revenues minus variable costs) as well as net income (variable revenues minus both variable and fixed costs) should be included. In addition, gross and net revenues after allowances for Medicare, Medicaid, and managed care adjustments must be identified.

PATIENT MIX

The operating costs associated with a subacute care service are highly dependent on the types of patients being treated as well as the focus of the service (subacute care medical and subacute care rehabilitation). Typical subacute care medical patients would include diagnoses related to pneumonia, shock, cardiac, pulmonary, oncology, and metabolic disorders. These types of patients tend to require extensive nursing care and make minimal use of ancillary services such as physical therapy and speech therapy. Nursing care hours for this patient population tend to be higher than for subacute care rehabilitation patients.

In contrast, subacute care rehabilitation patients can be categorized as having the potential for functional improvement. In addition, this patient population uses therapy as an integral part of the recovery process. Subacute care rehabilitation patients also have the ability to benefit from the rehabilitation model of care.

Typically, diagnoses related to neurologic and orthopedic ailments constitute the majority of cases for these patients. It is important to note that patients who are not usually associated with rehabilitation, including diagnoses reflecting neoplastic and pulmonary conditions, also find the rehabilitation model of care conducive to recovery.

Physical, occupational, and speech therapies constitute a significant portion of the rehabilitation model of care. Through this model, ancillary services augment the nursing care provided in a more interdisciplinary approach. From a reimbursement perspective, costs associated with ancillary services are outside the routine costs that are subject to limits.

STAFFING

Salaries, wages, and benefits account for a large portion of the operating expenses associated with any level of healthcare, and subacute care is no exception. Judicious use of nursing staff can help the program achieve the lower expenses that distinguish subacute from acute care. If acute staffing patterns are utilized, the cost of care would be prohibitively high, significantly exceeding Medicare routine cost limits. Traditional nursing home staffing would not be intense enough, however, to provide the higher level of care necessary in this patient population. The operating parameters that distinguish subacute care from traditional skilled nursing care are summarized in Exhibit 5.2.

EXHIBIT 5.2
COMPARISON OF SUBACUTE AND SKILLED CARE

Care Variable	Subacute Care	Skilled Nursing
Admission Criteria	Program-specific with clear outcome potential (orthopedic, neurology, cardiology, etc.)	Any diagnosis; Medicare skilled criteria
Average Length of Stay	7 to 21 days	90 to 180 days
Average Charge/Day	$300 to $700	$100 to $200
Physician Visit	Average 1 to 3 times per week; consulting	1 time per month; limited consulting
Nursing Care	4.5 to 8.0 hours direct care; high use of aides, nonprofessional	2.5 to 4.0 hours
Rehabilitation Therapies	Programmatic with specific goals; usually more than 1 hour per day	Limited in scope; usually less than 30 minutes per day

Typically, nursing services average 4.5 to 8.0 nursing hours but may be higher for ventilator dependent patients or other complex patients. The mix of the staff, that is, the ratio of licensed nurses to aides is 1:2 to 1:1, depending on the nature and needs of the patient population being served. Other routine staff directly assigned to the service would include a medical director, an administrative director, a clinical coordinator, and other secretarial and administrative staff. Exhibit 5.3 presents a typical staffing pattern, expressed as full time employees (FTEs), for a 20-bed subacute care program with an average daily census of 15 to 18 patients.

When projecting the performance of a subacute care service, the planner must group the costs on a routine, ancillary, and capital cost basis in order to calculate costs in a manner consistent with Medicare reimbursement limits. Capital costs for distinct-part SNF units continue to be reimbursed on a cost basis (unlike the prospective payment basis for acute care services).

EXHIBIT 5.3
PROJECTED SUBACUTE UNIT STAFFING FOR TYPICAL 20-BED UNIT

	FTEs
NURSING	
RNs	4.50
LPNs/LVNs	4.50
CNAs	6.50
Subtotal	15.50
THERAPY	
Physical Therapy	1.20
Physical Therapy Assistant	1.50
Occupational Therapy	1.20
Speech	1.00
Subtotal	4.90
OTHER ROUTINE COST CENTER	
Medical Director	.20
Director of Nursing	.50
Administrator	.50
Case Manager	1.00
Secretary	.50
Business Office	.50
Subtotal	3.20
TOTAL ROUTINE FTEs	23.60

ROUTINE SERVICES

Routine services represent the costs associated with operating the service and include room and board along with nonrevenue-producing cost centers of administrative and general, plant operations, laundry and linen services, housekeeping, dietary, cafeteria, nursing administration, medical records, and social services. Routine services for subacute care include additional staffing (both administrative and clinical), nonchargeable supplies, minor equipment, and maintenance contracts. Some subacute care services incur additional reimbursable expenses within management contract fees. Providers should be aware that Medicare recognizes only the actual cost of providing the management service from one related party to another. Routine costs, including direct salaries, wages, and benefits for the nursing and managerial staff, represent those costs subject to reimbursement limits for skilled nursing providers.

ANCILLARY SERVICES

The ancillary services most often utilized by subacute care patients are the physical, occupational, and speech therapy departments as well as diagnostic radiology, laboratory, intravenous therapy, respiratory therapy, and supplies and pharmacy charged to patients. As discussed previously, the level of ancillary utilization will vary by type of patient. In general, subacute care medical patients can require between 0.25 and 0.5 hours of rehabilitation treatment per patient day, whereas subacute care rehabilitation patients may need one to three hours of therapy service per patient day. Physical therapists, occupational therapists, speech therapists, and their aides represent the vast majority of ancillary service revenues and expenses.

To determine the estimated costs and revenues for therapy and ancillary services, the expected charges can be applied to the facility cost-to-charge ratio to determine fully allocated costs. As with routine services, ancillary expenses are calculated on an incremental basis. In addition to calculating the fully allocated ancillary service expenses, *reimbursable overhead* for therapy services must also be included in the calculations. Reimbursable overhead is the amount in addition to the actual salaries and wages of the therapy providers that is incurred as a result of operating the unit. Exhibit 5.4 represents hospital-based subacute care ancillary charges, fully allocated and incremental expenses for a typical general subacute care service. As noted, incremental expenses represent 36% of fully allocated expenses.

COST STRUCTURES

Capital Costs

Capital costs consist of depreciation and interest on existing assets as well as additional assets necessary for the facility or unit. These capital costs could relate to site acquisition or development, architectural and consulting fees, actual construction or renovation expenses, and costs associated with fixed and movable equipment. In addition, capitalized interest and other financing expenses can be included. For the retrofitting to convert existing acute care beds to subacute care, reimbursable capital costs will be determined through the

EXHIBIT 5.4
SUBACUTE ANCILLARY CHARGES AND EXPENSES

	Projected Charges per Patient Day	Fully Allocated Projected Costs per Patient Day	Incremental Projected Costs per Patient Day
Radiology	$ 16.00	$ 8.79	$ 3.62
Laboratory	18.00	12.01	3.94
Respiratory	30.00	19.55	6.41
Therapy	100.00	55.73	25.51
EKG/EEG	1.00	0.58	0.29
Medical Supplies	30.00	29.62	7.98
Pharmacy	20.00	20.14	11.17
TOTALS	$215.00	$146.42	$58.92

* Average based on experience of more than 100 hospital-based providers.

Medicare cost allocation process. Specifically, any additional subacute care capital costs would be pooled with those of the existing facility and allocated on a square footage basis to the subacute care program. The potential effect of this is to shift existing capital dollars from DRG-based to cost-based centers.

Physical plant requirements, similar to patient care requirements for subacute care services, fall somewhere between those required for acute providers and traditional nursing home providers. If financially feasible, medical gases should be available in the patient rooms. In addition, activity areas and dining facilities must also be available to the patients. Sufficient therapy space in close proximity to the unit also needs to be included.

Depending on the nature of the construction involved (new construction versus renovation) as well as the type of facility (acute versus freestanding nursing home), capital costs can vary from retrofitting an existing unit to new construction. The expertise of an experienced architect specializing in health-care facilities is necessary to determine the magnitude of the construction costs.

As mentioned previously, reasonable and necessary capital costs are reimbursable by Medicare. To be reimbursable, any interest expense associated with a loan must be deemed necessary. If sufficient reserves are available to fund the project, the interest amount may not be an allowable reimbursable expense.

Movable equipment for subacute care services is not capital intensive and consists primarily of beds, furnishings, and therapy equipment.

Incremental Costs

Incremental costs are variable expenses that increase proportionally with each unit of service delivered. As utilization increases, associated expenses also increase but not in a linear fashion. The degree to which the costs vary differs by cost center, with some experiencing highly incremental costs whereas others vary minimally to volume changes. The cost relationships to revenue chart in Exhibit 5.5 illustrates the incremental concept, whereby as volume increases, revenue and total cost increase while fixed costs remain constant. Incremental costs represent the difference between fixed costs and total costs.

EXHIBIT 5.5
COST RELATIONSHIPS TO REVENUE

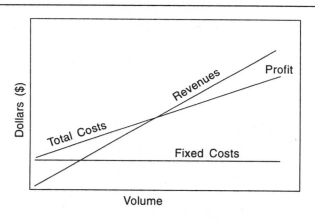

To further illustrate this point, the cost centers of laundry and plant maintenance can be reviewed. If the facility does not have a cost accounting system that identifies variable costs, the Medicare Cost Report can be used to approximate these relationships. By using Worksheet B, Part I (cost allocation of general services), the direct and indirect allocations can be determined. Worksheet B, Parts II and III, indicate the amount of old and new capital associated with each cost center. This amount must be subtracted because capital is a separately reimbursable expense. Worksheet B-1 indicates the cost allocation factor used by the various cost centers. Exhibit 5.6 shows the total costs, capital costs, and the direct costs for the laundry cost center. In addition, the allocation factor is also indicated as well as the per-unit-of-service expense.

EXHIBIT 5.6
COST ALLOCATION FOR LAUNDRY SERVICES

	Expense	*Per Unit of Service*
Direct		
Salaries	$ 628,369	$0.21
Other	915,339	0.31
Total Direct	1,543,708	0.52
Indirect Allocated		
Capital	637,958	0.22
Benefits	113,117	0.04
Administrative & General	298,491	0.10
Telephone	4,505	0.00
Purchasing	18,212	0.01
Maintenance	253,308	0.09
Total Indirect Allocated	1,325,591	0.45
Total	2,869,299	0.79
Total Less Capital	$2,231,341	$0.75

For the example for an average 500-bed hospital facility, total direct laundry expenses were $1,543,708 for laundry department salaries, wages, and supplies. The fully loaded expenses after all allocations were $2,869,299. Associated capital costs amounted to $637,958. The allocation factor used for laundry is pounds of laundry processed. In this example, 2,959,251 pounds of laundry were processed, which equates to $0.75 per pound fully allocated cost for laundry, compared to $0.52 per pound when looking at direct costs only.

For typical subacute care programs, the additional incremental laundry cost for each patient generally consists of some additional laundry supplies and labor. An increase of 5% in patient days would not increase laundry costs by a similar amount, because additional staffing most likely would not be necessary. In contrast to laundry, plant operations are nearly 100% fixed. Regardless of the census of the program, expenses associated with this cost center remain fairly constant. By differentiating between fully allocated and incremental costs, a projection of the service's contribution after incremental costs, but prior to fixed costs allocations, can be made.

Traditional program of service allocation factors include square footage, accumulated costs, pounds of laundry, and patient days. For purposes of projecting the cost allocated per patient per day to the cost center on the cost report can be used, on an adjusted basis, as a barometer of allocated subacute care expenses. In addition, for each cost center the percentage of incremental cost needs to be determined. This determination is most accurately made through facility-wide cost accounting systems. If the provider has access to Medicare Cost Report software, the facility-wide effect can also be assessed.

It is important to note that for acute providers a significant advantage exists by shifting costs from fixed payment sources to cost-based reimbursement sources.

REVENUES

Total gross revenues for subacute care can average between $300 and $700 per patient per day. Charges need to reflect the market conditions of the area, as well as the actual expenses incurred by the provider. Revenue sources are divided into room and board, therapies, and ancillaries, which typically account for 50%, 30%, and 20% of total revenues, respectively. These charges are then applied to the projected patient days to determine the amount of revenue the service can generate, prior to contractual adjustments.

MEDICARE REIMBURSEMENT

Contractual adjustments represent the difference between gross patient revenues and payments made by third parties, primarily the Medicare program, along with selected Medicaid, insurance companies, and managed care programs.

Medicare reimburses on a cost basis for subacute care services provided in SNF-licensed beds. The routine reimbursement amounts and adjustment factors are published by HCFA. The basis for routine reimbursement are the labor-related and nonlabor-related group limits. These vary by skilled nursing facility designation, freestanding or hospital-based designation, and by location of the facility, that is, whether it is in a metropolitan or rural location. Currently, hospital-based facilities receive higher reimbursement, as do facilities in urban locations.

To determine a facility's labor-adjusted limit, the labor-related component is multiplied by the wage index for the area. Then the nonlabor-related component and the nursing home reform and Occupational Safety and Health Act (OSHA) per diem add-ons are figured in to determine the total routine cost limit. The resulting number is then adjusted by a factor to account for inflation from the time that these limits were published (October 1992) to the date on which the facility is expected to open. In addition to the routine reimbursement, Medicare reimburses ancillary services on a reasonable cost basis. Depending on the level of utilization, reimbursement for ancillary services can approximate the amount of routine cost reimbursement.

Medicare recognizes that, during initiation of a new service, a buildup of utilization occurs. Regardless of utilization, fixed costs of operating would be experienced. As a result, a facility would experience high routine costs per patient day if utilization is low. Therefore, facilities initiating new Medicare SNF services are granted an exemption to these routine limits for the first three years of operation. During that period, providers are reimbursed for the actual costs incurred.

An additional alternative for relief from the routine cost limits also exists. An exception is granted to a provider that can document legitimate justification for exceeding the limits. Justifications deemed acceptable by Medicare include extraordinary circumstances (such as natural disasters), provision of care to areas with fluctuating populations, and provision of atypical care.

Atypical care includes care to patients who require a considerable amount of nursing and rehabilitation care, patients with more serious illness than generally seen in SNF beds, a high proportion of Medicare utilization, and patients with very short lengths of stay. Medicare recognized that all of these factors would increase a provider's routine costs. The routine limit is based on an aggregate average of all providers; therefore these atypical services would not be represented by the average cost.

MEDICAID REIMBURSEMENT

Medicaid reimbursement for subacute care varies by state. Many states pay an all-inclusive per diem or per case amount, regardless of the ancillary utilization. It is important to note that an additional source of reimbursement for elderly Medicaid patients is Medicare Part B coverage. If the patient is eligible, ancillary utilization in a subacute care program is reimbursable under Part B coverage.

Since many states do not separately recognize Medicaid reimbursement for subacute care services, it is important to have admission and discharge criteria that distinguish subacute care patients from nursing home patients. Specifically, problems can occur when Medicaid patients admitted for subacute care recover to the point that only traditional skilled services are required. In some instances, the state Medicaid department may not wish to facilitate a transfer if there is no recognized difference in Medicaid subacute care and traditional programs.

MANAGED CARE AND OTHER PAYOR REIMBURSEMENT

Reimbursement for nongovernmental payors can be structured as a percentage of charges, on a per diem rate, on a per case basis, on a capitated basis, or a combination of any of the aforementioned. To most accurately project net payments, each payor should be addressed separately. However, a reasonable estimate can be calculated using a percentage of the gross charges.

CONTRIBUTION ANALYSIS

A subacute care service's financial contribution is the difference between variable revenue and variable expenses. In instances where an existing facility has excess or unused space, the contribution represents the incremental amount the service contributes toward covering fixed overhead. When space is converted, the contribution of the previous service should be compared to the new service to determine true financial benefit.

Net income is calculated by subtracting all variable and fixed expenses from subacute care revenues and represents income for the service after all indirect allocations are made. The majority of an acute provider's revenues come from prospective payment systems. By shifting certain fixed expenses to a developed subacute care program, a provider achieves cost-based reimbursement while having no reduction of prospective revenues. When determining the overall impact of providing subacute care, the effect on the remaining operations should also be considered. The decreased length of stay and reduced resource consumption should also be analyzed.

SUMMARY

Overall, subacute care presents opportunities for providers to extend the continuum of care that they currently offer while enhancing financial performance. Providers need to be aware that, as with any potential opportunity, challenges also exist. The challenges include the few industry standards relative to reimbursement issues, competition among providers at various levels of the continuum of care, and ability to recruit and retain staff with sufficient experience and expertise to care for these patients cost effectively. Regardless of these challenges, subacute care remains an attractive financial development opportunity for many providers.

CHAPTER 6

The Consumer

"The aim in building sound attitudes about health is to educate people."

— Dr. Benjamin Spock

Whether a subacute care facility is successful depends a great deal on the consumer. The consumer is the end-user of the services, specifically, the patients treated. Unlike other markets, the consumer is influenced by many issues in deciding on healthcare services. Some experts believe that in no other industry do external forces influence decisions to a greater degree than in the healthcare services industry.

If the consumer is not the primary source when it comes to healthcare decisions, it is for the following main reasons:

- The consumer doesn't understand the options
- The consumer is not the payment source
- The consumer's family is often involved
- The consumer does not have confidence in the care givers.
- The consumer is physically incapable of making the decision.

In a relatively newer service such as subacute care, the consumer will depend more on outside influences.

The most compelling and overwhelming issue that the subacute care industry must address is the consumer's apprehension and fears of being moved about while trying to recuperate. That fear increases if it involves a move outside the traditional hospital setting for care, especially, it seems, if it involves transferring a patient to a nursing home.

The term *nursing home* may conjure up some negative thoughts. This fear of nursing homes by consumers is, and will remain, the biggest hurdle to accepting their services. Just like the root of most fears, these negative images are founded in ignorance or in fear of changing one's environment.

It is important for the consumer to understand and want to utilize subacute care programs. Healthcare companies build clinics, hospitals, laboratories, nursing homes, imaging centers, surgery centers, physician office buildings, and other facilities at an unbelievable rate without determining the needs of the community or assessing demographics and trends. The amazing fact is, typically these competitive entities are built adjacent to or in the same neighborhood as the existing ones. Consumers must be educated regarding what subacute care is and how it can benefit them.

CHANGING TIMES FOR HEALTHCARE CONSUMERS

The consumer of healthcare services is changing in many ways; some are remarkable. The consumer's level of knowledge has increased tremendously about healthcare services and costs. Much of this is due to the focus on healthcare reform. For the first time ever, customers are questioning the need for various medical services and products. Age, income levels, and attitudes are changing and in the future will change even more and faster. The consumer is becoming less trusting of the healthcare profession. The infallible physician is a thing of the past. The consumer is realizing that the physician is human and occasionally makes mistakes. There are options for care that the physician may not be aware of or agree with, and for these reasons may never offer to the consumer. As consumers educate themselves, we will see a change in relation to the manner in which physicians practice and interact with their patients.

Consumers will exercise greater influence in choosing healthcare providers because they will have

- The ability to choose providers
- Exposure to marketing and advertising by the healthcare industry
- A greater awareness of more and different options
- Control of their own healthcare funds

Subacute care providers must realize that this change is beginning to happen and will have a major impact.

Currently, three separate demographic phenomena are occurring simultaneously. Each one separately would have an effect, and together they will forever change all industries, including healthcare and, specifically, subacute care. These are:

1. The senior strength—Americans have longer life expectancies than ever before; the elderly are more active, healthier, wealthier, and more influential than any previous generation has been.

2. A decreased birth rate—For two decades the birth rate has been at a very low level. The population of older adults is not being matched by any increases in births.

3. Baby boomers becoming "grayby boomers"—The baby boomer generation is aging and a large percentage are already into their 40s and starting into their 50s.

These three factors are combining to produce a historic shift in the method by which all industries deliver products and services. Since more attention is given to health-related issues and the improvements in healthcare services that add years to our lives and improve quality, this industry will become a significant focus.

At each stage of their lives the demands and desires of baby boomers have and will become the dominant focus of business. Some examples are:

* When the baby boomers arrived, the diaper industry developed exponentially.

* The baby food industry went from 270 million jars sold in 1940 to 1.5 billion by 1953.

* As boomers grew, so did the pediatric medical establishment.

* Related industries also expanded, such as paper products.

* More elementary schools were built in 1957 than any year prior or since, and in 1967 more high schools than ever before or since.

* The central theme for advertisers was the child and the family.

* Fast food became a staple of the American diet.

* In 1964 alone, teenagers spent $12 billion and their parents another $13 billion on them.

* College enrollment went from 3.2 million in 1965 to 9 million in 1975.

* In the early 1980s, as boomers moved through their thirties, day care and personal financial industries quadrupled.

* In the 1990s, as boomers move toward their 50s and 60s, the future of healthcare is among the main concerns.

Following are some additional facts that impact healthcare services of the future:

- In recent decades the average life expectancy has increased by 28 years.
- The life expectancy by 2020 for men will be 86 years and 91.5 years for women (The National Institute on Aging, a division of the National Institutes of Health).
- 80%–85% of all Americans will live past 65 years of age.
- In July of 1983, the number of Americans over 65 surpassed the number of teenagers.
- Almost one-half of the 65+ population is over 75 and the 85+ group is the fastest growing segment. Today, it numbers 3.6 million, and by 2050 will exceed 20 million.
- By the turn of the century, there will be more than 35 million Americans over 65 years of age. This equates to one-seventh of the entire population. And by 2040 the National Institutes of Health projects that 90 million Americans will be over 65 years old.

This massive and influential group will not wait around until they celebrate their sixty-fifth birthdays to confront their healthcare needs. The time to act is within the next five years, and the resulting decisions will have an impact on all future generations.

In fact, boomers are dealing with healthcare decisions today—those of their parents. Already, one-third of the estimated $500 billion spent on healthcare goes to the 12% of the population over 65.

Consumers do and will have the financial leverage to pull American businesses in new directions. As a matter of fact Americans over 50, though currently only 25% of the population, have a combined annual personal income of over $800 billion and control over 70% of the total net worth of U.S. households estimated to be nearly $7 trillion in wealth.

They own 77% of all financial assets and 80% of all monies in U.S. savings and loan institutions. These consumers purchase 80% of all luxury travel, spend more on healthcare than any other group, and account for 40% of total consumer demand. This group is more prudent and quality conscious than any other group.

BENEFITS TO CONSUMERS

Subacute care can offer the consumer many benefits but as with other service areas, the benefits may not always be apparent.

For the consumer to understand the benefits of subacute care, it is imperative to state those benefits in a clear and concise document. One way to ensure that the consumer receives the message clearly, can internalize it, and can share it with others is to state the features along with associated benefits; this is known as *consumer adoption*. This is based on the fact that consumers will most likely be making the decision with others and/or for others.

Insurance Benefits Extended
In many instances, if the patient is moved from an acute care hospital bed to a subacute bed, hospital days are saved. Saving hospital days, which are most often limited to a set number of days per year, would enable a patient to be admitted or readmitted to an acute care hospital with less anxiety about coverage and payment. Subacute care will extend benefits because of their lower rates and will not exhaust the cap on the policy as quickly.

Reduction in Premiums
The consumer who reduces hospital days may help reduce or at least maintain his or her insurance premiums for the future years. This often occurs because health plans establish and review rates of insured groups based on expenditures, and the largest expenditure is hospital days. Additionally, because the total healthcare dollars spent will be reduced by utilizing subacute care services, adjustments to reduce rates may be warranted. This scenario is similar to the auto insurance "good driver" program that offers lower rates for minimizing the utilization of services and claims.

Since a majority of healthcare services are reimbursed through indemnity insurance plans or HMOs, deductibles or copayments in the form of out-of-pocket expenses are factors to consider. Subacute care services, because they are less costly than acute care hospital services, may reduce these expenses, as Exhibits 6.1 and 6.2 illustrate. As you will see, because the total cost for Patient X who chose to use subacute care services was less, the copayment was significantly lower. In both instances a significant amount of out-of-pocket expenses are saved. Additionally, a reduction in total expenditures may help reduce future years' premiums.

Other benefits include a greater accessibility for families because most subacute care programs have flexible visiting hours. Thus, family, friends, and loved ones can visit at their convenience. Additionally, most subacute care programs do not have limited parking and do not charge for using their parking lots. Another benefit is that subacute care programs encourage a personal physician to visit and provide treatment. Since hospitals treat higher acuity patients with a variety of diseases, the infection rate tends to be higher than that found in a subacute care center.

Often patients in a hospital are moved from unit to unit and room to room depending on their status. It is not unusual for the hospital personnel to change,

---

EXHIBIT 6.1
DEMONSTRATED SAVINGS IN COPAYMENTS

Patient O	*Patient X*
O has orthopedic surgery in an acute care hospital; remains in the hospital postsurgically for recuperation, including medication and rehabilitation therapy for seven days	**X** has orthopedic surgery in an acute care hospital; is transferred to a subacute center with orders for medication and rehabilitation after two days of postoperative care

Hospital days: 7	Hospital days: 2
Cost per day: $1,000	Cost per day: $1,000
	Subacute days: 5
	Subacute (per day): $500
	2-day acute stay: $2,000 5-day subacute stay: $2,500
Total cost = $7,000	Total cost = $4,500
Copayment (20%) = $1,400	Copayment (20%) = $900

Note: This does not factor in surgical costs or physician costs.

---

especially nurses and nursing assistants who provide the majority of care. These staff members are often assigned to other units or are temporary staff from an agency. Patients rarely see the same care giver if they change units or floors in a hospital. In subacute programs, patients are not shuffled to various units and the care givers are typically the same day after day. This allows the patient and care giver to interact and for the care giver to better learn and meet the patient's needs. It is frightening to be ill, and constantly changing personnel can raise anxiety and increase mistakes as familiarity with the patient decreases. Consistency in care can promote recovery.

In most subacute care programs the staff becomes more familiar with the patient and often the patient-to-staff ratio is greater. This encourages an environment where patients receive training on the use of equipment such as IV pumps, nebulizers, and walkers. Patients are also educated regarding medication regimens such as insulin, and adaptive supplies such as urinary collection devices

---

EXHIBIT 6.2
ILLUSTRATED SAVINGS IN DEDUCTIBLES

Patient O	*Patient X*
O has an asthmatic episode that requires a visit to the emergency room; is evaluated and will require 2 days of observation and respiratory care; is admitted to the hospital for these services	X has an asthmatic episode that requires a visit to the emergency room; is evaluated and will require 2 days of observation and respiratory care; patient is admitted to a subacute care program
Emergency services: $500	Emergency services: $500
Hospital days: 2	Hospital days: 0
Cost per day: $1,000	Cost per day: 0
	Subacute days: 2
	Subacute cost = $500
Total cost = $2,500	Total cost = $1,500
Patient O's Deductible—$2,500	Patient X's Deductible—$2,500
Due from Patient O = $2,500	Due from Patient X = $1,500

This does not include physician costs and may not include some services.

---

and necessary equipment. Additionally, subacute care team members train family and other care givers to assist the patient.

CHOOSING SUBACUTE CARE SERVICES

Since subacute care is just becoming known to consumers, the degree of familiarity with subacute care services is minimal but will increase significantly in the future. Consumers often consider many factors; their healthcare knowledge base and the degree to which they are influenced by external forces will help determine their decisions. Consumers should ask many questions to learn as much about their care as possible, such as the following:

General Questions:

- When was the last survey and what were the results?
- Were any deficiencies cited? How have they been rectified?

- How many beds are available for the type of care needed?
- How many beds are available for Medicare patients? Medicaid? Insurance?
- Are private rooms available?
- What special services are available?
  - Respiratory Care
  - Ventilator
  - Advanced Therapies
  - Dialysis
  - Postoperative
  - TPN
  - Home Care Training
  - Hospice Services
  - Cardiac Care
  - Infusion
  - Oncology Services
  - Other (specific patient need):_____
- Is transportation provided to/from physicians' offices? The hospital? Clinics?
- If the patient needs hospitalization, which hospital(s) would be used?
- What constitutes an emergency?
- Is the program part of a network of services? Other facilities?
- Is it permissible to go off grounds for short periods of time?
- Are visitors permitted to spend the night? (This may or may not be appropriate dependent on each situation.)
- Does the program have a case manager? What are his or her credentials? (A case manager is a healthcare professional, usually an RN, often certified, and should not be confused with the facility admission coordinator or other personnel.)
- Is there a financial relationship with a hospital?
- Who is the medical director?

Activities
- What types of activities are included?
- What are the visiting hours?

Nursing

- Who is the director of nursing?
- Who is the clinical nurse specific to the subacute care program?
- What is the patient-to-registered nurse ratio?
- How many patients does each nurses aide care for?
- How long has the average nurse been employed at the program?
- What are their backgrounds (acute care, rehabilitation, other)?
- Does the program use temporary or agency nurses?

Therapies

- What therapies are provided?
- Are they provided by employees or outside agencies?
- Are the therapy sessions one-on-one or group sessions?
- Are there licensed or registered therapists, that is, physical, occupational, speech, respiratory, recreational?
- Is there a gym and equipment?
- How many days per week are therapies offered?

Physician Services

- Who provides physician services?
- Is there a specialist to oversee programs?
- Can my physician or specialist treat me?
- How often will physicians visit?

Payment for Services

- Does the facility accept the patient's insurance?
- What is the rate for care?
- What is included and excluded in the rate (laundry, personal items, nonprescription medications, medical equipment, therapies, and so on)?
- What if the insurance coverage period lapses prior to discharge?
- Are TV and telephone included? If not, what is their cost?

Other

- ◆ How many patients with this condition have been treated in this program?
- ◆ What has been the outcome? Length of stay?
- ◆ How many other patients from the same insurance company have been admitted to this program in the past year?

If Services Are for a Child or Adolescent

- ◆ Is education provided?
- ◆ Are activities age appropriate?
- ◆ Is the staff trained specifically for pediatrics?
- ◆ Are the patients cared for in the same unit as adults?
- ◆ Can parents/family stay overnight with the child?

The consumer's predominant reason for choosing a facility will be dependent on recommendations by the various influencing parties, especially the physician, hospital social worker and/or discharge planner, insurance case manager, and family members. Exhibit 6.3 illustrates these influences.

THE APPROPRIATE TYPE OF CARE

Different types and levels of care are necessary for various patients. As with other healthcare services, subacute care is not always the optimal type of service. The appropriateness of using a subacute care program depends primarily on the following:

1. The clinical status of the patient
2. The subacute care program's capabilities
3. Whether the subacute care program's services provide optimal benefits for the patient
4. Whether subacute services are reimbursed via the consumer's insurance or other source

A subacute care program is appropriate and should be considered if based on positive responses to the above and the consumer's comfort level. If the subacute care program can reduce the patient's length of stay in the hospital, assist the patient in getting home at an earlier date, reduce the chance of infection, and improve the outcome, it would be prudent for the consumer to consider this option.

EXHIBIT 6.3
INFORMATION SOURCES INFLUENCING A SUBACUTE CARE DECISION

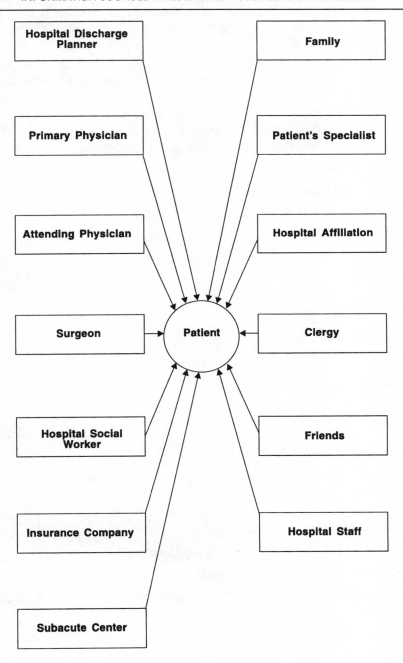

One example of a situation in which subacute services are extremely appropriate and beneficial, and have been utilized for some time with positive results is rehabilitation. This program is designed for patients who require extensive rehabilitation care yet cannot benefit from or tolerate a minimum of three hours of therapy in an acute rehabilitation setting. The rehabilitation patient does not, and should not, remain in the hospital and is inappropriate for acute rehabilitation. Such patients may even regress because they cannot handle an intensive level of therapies. Subacute care services can bridge that gap and prepare a patient for the rehabilitation hospital or, many times, progress from the subacute care program directly to home.

Other examples of specific subacute care programs are:

- Alzheimer's care
- Dialysis care
- Enteral services
- HIV/AIDS care
- Infusion care
- Nutritional services
- Oncology services
- Pediatric care
- Pre- and postoperative care
- Pre- and post-transplant care
- Rehabilitation care
- Respite care
- Tracheostomy care
- Wound care

It is important to note that not all subacute care centers have the same capabilities and should only be considered if they provide the appropriate level of care. For instance, not all subacute care units provide ventilator care, transplant care, or pediatric services.

Subacute care program settings are not appropriate if:

- The patient is clinically unstable
- Frequent diagnostic testing is required and cannot be provided at the facility
- The patient can be cared for effectively at home or in a traditional skilled nursing facility

◆ Surgery is required.

The rule of thumb to use when deciding whether subacute care services are appropriate or not is: is it the most appropriate setting? If so, subacute care services must be considered.

TOUR OF THE FACILITY

No verbal information can substitute for a personal visit to the subacute care program. During the visit, observe the patients, staff, and the general condition of the facility. Assess whether the facility is clean and well-staffed, and whether the patients appear comfortable and content. Talk with as many people as possible to get their opinions of the personnel and conditions of the program. The staff's responsiveness to the needs of the patients should help to determine whether the patients are provided adequate privacy. Discern whether staff, especially those providing hands-on services, appear competent and content with their jobs.

Attempt to speak with at least three staff members. The administrator should be able to answer questions immediately or get an answer quickly. The director of nursing should be able to address the specific needs of the patient. If a special diet is required, the patient or family should speak with the facility's dietitian. A case manager should be available to discuss program issues.

Family members can and should play an active role in patient care. They should talk with the staff about how they can be involved in the care plan, potential rehabilitation, and equipment training. Family members should be invited to attend care plan meetings to learn and assist in the process. Patients whose families are actively involved in the care have a better opportunity for rehabilitation and recovery, which could result in a more timely discharge to home.

PROVIDER CONSIDERATIONS

Consumers buy products or services for a variety of reasons. A typical consumer goes through a series of stages in the process of adopting a new service or product, such as subacute care. They are:

1. *Awareness*—The consumer becomes cognizant of the innovation but lacks information.

2. *Interest*—The consumer is stimulated to seek information about the innovation.

3. *Evaluation*—The consumer considers whether it would make sense to try the innovation.

4. *Trial*—The consumer tries the innovation and estimates its utility.

5. *Adoption*—The consumer decides whether or not to make use of the innovation in the future and influences others with his or her opinion.

Consumer considerations when planning to purchase the services are:

+ *A Solution*—Does the service offer a solution to the need?

+ *Features and Benefits*—Does the consumer understand the benefits of the services? Typically, a consumer requires at least one or two benefits.

+ *Comfort and Convenience*—Do the services offer these qualities?

+ *Performance*—Does the facility have outcome data and case studies to review? Some prospects will buy what they perceive as the best product.

+ *Image*—Does the service reinforce the image consumers have of themselves?

+ *Quality*—Does it appear to be a quality service?

+ *Price*—Is it priced appropriately? Some consumers will demand low prices.

+ *Relationship*—Does the facility have name recognition and/or affiliations with hospitals and physicians?

+ *Selection*—Does the facility offer a wide range of services? Offering home care or nursing services to patients going home may help increase acceptance and be an added benefit.

+ *Location*—Is the facility accessible, and in what kind of neighborhood is it located?

+ *Tradition*—Do consumers know others who have used the program?

+ *Innovation*—Is it a new service offering? New services attract new business.

+ *Emotions*—Does the purchase make consumers feel good?

One major factor to keep in mind regarding the decision to utilize subacute care services is that a decision may be required quickly and may be for a short-term stay. In these cases, which are increasingly common, recommendations from the physician, case manager, and social worker will prevail.

There will be an awakened interest in the consumer market as the shift in supply and demand relationships increases, competition increases, and the consumer begins to have more influence in the choice of providers of subacute care.

These issues will force industries to design and sell products and services that meet consumer desires and demands. Physicians, case managers, social workers, and others will most likely always play a large role in influencing decisions regarding the use of subacute care services and the choice of providers. Tailoring plans to include the consumer's perspective will enable a subacute care program to keep pace and to present subacute services as a viable option.

SUMMARY

With the impact of changes in healthcare, the future growth of managed care, the aging of the baby boomers, and the size and growth potential of healthcare (specifically, this segment called subacute care), the consumer will be an increasingly influential decision maker.

The healthcare industry needs to begin to educate everyone about subacute care, not just for the potential business, but for the population as a whole. The key to the success of a new industry is how well it is understood and the benefits attained by its usage. Consumers must take an active role in the healthcare delivery process. The more informed people are, the more intelligent their choices. This is part of our future health.

CHAPTER 7

Managed Care

"Books are too important to be locked away in a barn. Books should be shared by everyone."
— Dr. Quinn, Medicine Woman

Managed care has become the dominant force in the healthcare marketplace for payors and providers alike, and subacute care is no exception. Increasing demand for cost-effective care has driven managed care providers to organize as integrated healthcare delivery systems. This will allow them to more efficiently coordinate the delivery of a continuum of care ranging from acute hospital and professional medical to subacute care, ancillary services, and home care.

FUNDAMENTAL CHANGES IN THE HEALTHCARE MARKETPLACE

The marketplace for the provision of healthcare services recently has experienced dramatic and fundamental changes. Payment methods have shifted from fee-for-service models to at-risk financial arrangements. These at-risk arrangements establish a fixed amount of payment irrespective of the cost or volume of services rendered. Quickly becoming a prevalent at-risk model is capitation payment, in which providers receive on a monthly basis a predesignated fee from a health plan or other payor for each patient enrollee assigned to the provider. This is in exchange for furnishing agreed-upon healthcare services for those patients. In this environment, the providers' focus has shifted from generating more services to capturing, retaining, and managing covered patient lives; the fewer services provided, the greater the profit.

In addition, under the managed care "gatekeeper" concept, primary care physicians have achieved greater prominence because they exercise substantial control over patient flow and, therefore, the healthcare dollar. It is likely that we will continue to see more primary care physicians and fewer specialists. Large primary care-based physician organizations have grown rapidly in recent years to the point where they oversee several hundred thousand patient lives within their systems. These medical groups will continue to expand and some will become publicly traded securities on the stock exchange. Many of these medical groups own or have relationships with hospitals. Some are part of large regional or national systems. Others are the managers in a managed care system.

Sophisticated providers possessing large amounts of capital—historically hospitals in the healthcare industry—now seek to direct more of their capital away from building and equipment to accumulating coverage of patient lives, primary care-based networks, and management resources.

An essential element of the healthcare process is the cooperation of providers, payors, consumers, and users of medical care. It does little good for organizations and individuals to have viewed their new partners-to-be as competitors or "the problem." To make subacute services and other managed care solutions really work most effectively, it is important for providers to understand the payor's view of the changing world of managed care and the repercussions of healthcare reform.

Healthcare providers must be cognizant of the payor's perspective and critical needs in the future healthcare market. It goes further than simply a knowledge of business practices. We must honestly value the vision, goals, and objectives of others. To do this, providers should review how payors have historically approached the business and the hurdles, both internal and external, they must overcome. It is easier to consider when the discussion is divided into the following stages: history, maturity/development, and the future.

HISTORY

Until the 1980s, insurance companies didn't consider the healthcare business to be in a crisis mode. Certainly customers were anything but happy about the large increases in healthcare plan costs, but although customer relations were severely strained, insurance companies learned ways by which to weather this concern. At this point, there were really no other options for large employers and other sponsors of medical plans. The insurance companies had long since exited the insurance business for large and medium-sized employee groups. They were serving as claim administrators and providing a form of utilization review through reasonable and customary payment guidelines. Further, they were compensated on a basis that measured the amount or number of claim payments. It was apparent that neither method provided an incentive to the insurance company to really manage costs and, therefore, the increase in costs continued to escalate.

Why did this happen? One answer could be that insurance companies were run by traditional managers whose training and expertise was often actuarially based. The indemnity business was one the insurance companies really knew and many had prospered in this arena. They were satisfied with conditions and were dragged, kicking and screaming, into managed care.

In the 1920s a prepaid health plan was developed for workers in Los Angeles, California. Later the parameters of this plan were adopted by E. F. Kaiser for his construction employees. Of course today, everyone is aware of Kaiser as a prime example of managed care. Many years later a number of health maintenance organization (HMOs), preferred provider organization (PPOs), and point of service (POS) plans began to develop.

Although insurance companies had been discussing managed care for years, they often were reacting to a perception that it would be years before 75% of the health coverage would be managed care. This continued to be just a prediction of things to come and an issue that required noting in a business plan. Throughout the late seventies and early eighties a limited number of these types of organizations formed mostly on a local or state level.

This issue of a future threat of managed care was exploded in the 1980s when Cigna boldly offered a point of service plan with guarantees against future cost increases to Allied Signal. This was a wake-up call to all insurance companies, and the illusive managed care model was replaced by an immediate concern.

Customers, through employee benefit consultants, suddenly demanded change, and insurance companies were required to provide healthcare networks with discounts to fill a need. This move surprised many insurance companies that, although they had been hearing for years about managed care and HMO growth being right around the corner, had decided not to invest in building networks and spending the millions of dollars necessary to compete.

By 1990, a handful had committed to participate on a full managed care basis. Whereas previously 10 to 15 insurance companies had provided the market for national contracts, in the early 1990s the number of competitors decreased. These companies desired the easy approach to providing a managed care alternative to their customers—easy in the sense that they quickly started building networks. These were not necessarily quality networks, but just networks. As a result, network development often took place in the wrong markets at the wrong price, involving the wrong providers. There were few controls and even fewer clear examples of improved results.

In those early days, the employee benefit consultants who controlled placement of insurance contracts did not know how to adequately measure the value of networks or even to compare networks. The emphasis remained on claim office productivity rather than examining network management or member service.

The quality of the network was immeasurable, so the insurance companies raced around the country signing agreements, normally PPOs, with almost any provider who would listen. The promise of volume in the form of increased patient flow was seductive to providers in exchange for a discounted price. Networks rarely had meaningful, documented results; therefore, no credit for these network results fell to the bottom line.

In the early years of managed care, when a company was successful in obtaining a new customer, it often disappeared from the marketplace for months at a time. The successful insurance company's resources were diverted to developing the network promised to a specific client. It was easy to track sales results by identifying who failed to bid on the next major client that came to market.

Meanwhile, because the carrier management personnel did not see any real improvement in their efforts to lower costs or even reduce the rate of increase, they continued to resist deeper discounts or guarantees of network performance to plan sponsors. The insurance companies had built what they thought was the answer to managed care, but the volume just did not develop.

Insurance companies' first attempt to solve the problem was to return to providers and try to convince them that deeper provider discounts should be offered. The providers felt that the insurance companies' failure to deliver the originally promised volume certainly did not justify further discounts. In general, the providers were unwilling to grant additional discounts based on new insurance company projections.

Although the Allied Signal guarantees by Cigna were not clearly understood in the industry, general insurers followed suit and 1990 and 1991 were years spent designing some pricing mechanism that would at least provide a shared risk. The insurance companies were back in the guarantee business.

MATURING/DEVELOPING

At this point, the large insurance companies took steps to

- ◆ Be more market responsive
- ◆ Develop new management
- ◆ Stop acting as uninvolved third parties
- ◆ Begin to make real guarantees

- Respond to customer demands
- Work closely with HMOs and other managed care organizations

Many partnerships were formed and insurance companies rapidly began to aggressively acquire HMOs. The goal was to reduce the rates at which healthcare costs were increasing from the 30% level to single-digit numbers. It worked. According to the U.S. Bureau of Labor Statistics, medical costs in general increased at only 5.4% in 1993, the smallest advance in 20 years. According to a survey conducted by the New York-based consulting firm of Foster Higgins, America's employers managed to maintain health benefit cost increases at single-digit percentages. Costs of health benefits rose only 8% in 1993. Although Foster Higgins indicated that several factors contributed to this slowdown, the key seemed to be the continued migration of employees out of fee-for-service indemnity plans and into managed care alternatives.

Although the industry experienced a reduction in the amount rates increase, customers continue to experience frustration because they believe that today's cost increases are still too high. Although the market is progressing to managed care at a rapid pace, even without healthcare reform, there is little doubt that President Clinton's efforts in 1993 to encourage the national healthcare debate spurred providers to take a self-policing action. It also forced many employee benefit plans to make difficult choices regarding managed care.

However, one problem remains. Employers are still the major buyers of healthcare insurance. Although employers remain displeased with the cost increases, in most cases a very small share of this rapid increase has been passed on to employees, either through higher prices or lower benefits. In fact, the $100 deductible and 80% coinsurance that was a standard benefit 30 years ago still exists in many employee benefit plans. Certainly deductibles have been raised and out-of-network usage has changed the coinsurance percentage, but basically only some employees have felt the full burden of rising costs.

Part of the solution, which payors and providers both must assist to implement, is a better program of educating the insured. This would include explaining the real cost of insurance and the benefits available.

THE FUTURE

With this scenario as a backdrop, where are we today? What must a subacute care provider do to make itself attractive to a payor? The answer lies in both the organization of the entity and how that entity sees itself and delivers coverage. In order to become an attractive partner, the management should review and, where necessary, implement the following five items:

1. Ensure that the provider is prepared to move effectively into targeted markets with flexibility and efficiency
   - Communicate a clearly defined vision/mission/strategy to ensure alignment of direction among all associates
   - Develop knowledge and skills of associates at all levels and locations
   - Develop an infrastructure to respond to national and regional payors/employers/consultants/case managers
   - Facilitate a culture change that supports increased flexibility, enabling an adjustment to appropriate managed care goals, at the same time maintaining a focus on census and budget
2. Position themselves as a key constituent within the new healthcare delivery system by cultivating business relationships with the following:
   - Payors at all levels
   - Large employers
   - Employee benefit consultants at national and regional levels
   - Buying coalitions that represent government associations and small business
3. Develop strategic alliances that will allow them to
   - Expand subacute services
   - Use a national and/or regional referral service plan
   - Capture the post-acute care market as it matures
4. Establish monitoring systems that allow them to
   - Track and respond to national and/or regional trends and opportunities
   - Develop a system to track and respond to government legislative activity
   - Monitor progress in responding to managed care growth and opportunities
   - Compare individual results to industry benchmarks
5. Provide the systems and medical protocols necessary to maintain its position
   - Establish a system to capture data critical to market assessment and planning
   - Provide outcome data documenting medical results
   - Illustrate quality of care and services provided in individual facilities

One of the necessary items for a good working relationship with a payor is the subacute care provider's understanding of the healthcare reform issues that are important to payors. It is not enough to know whether payors are for or against certain issues or what their lobbying positions were in the recent debate. Rather, providers should try to understand why payors took their positions and why certain issues were viewed as critical. The key issues viewed as critical by insurance companies should be mutually explored. A few of them are:

+ Backstop on spending
+ Universal coverage
+ Standard benefits packages
+ Any willing provider
+ Single-payor systems
+ Community rating
+ Antitrust immunity

Any potential subacute care provider should know exactlv what the insurance company's position is on each issue and, just as importantly, why.

When most insurance companies and providers think about managed care, a handful of concerns must be addressed. Typical questions are:

1. What does subacute care mean to different entities?
2. How did we arrive where we are and where do we think we are going?
3. What is our role?
4. Can we make a difference?
5. What are the positives and negatives for our participation in such a program?

Often organizations leap to the conclusion that creating a subacute care program is the perfect plan for them without first going through an exercise of answering the previous questions or establishing the critical success factors needed to sustain a viable subacute care program.

Although these are issues to be reviewed before the provider decides to develop subacute care programs, it is important to discover what is really important to the payors. Find out key issues that need to be addressed when considering opening a subacute care program in a facility. Experience has shown that there are some basic issues to be considered in terms of payors:

+ Price
+ Accessibility

- Capabilities
- Experience
- Quality care
- Communication
- Program case management with individual flexibility
- Outcome-focused programs

Critical success factors are:
- Know who you are
- Learn your strengths and weaknesses
- Assess your potential
- Prepare to move into markets with flexibility and efficiency
- Develop strategic alliances
- Establish monitoring systems to track
- Measure managed care activity
- Monitor legislative actions
- Track industry trends
- Establish medical protocols
- Collect and develop outcome benchmarks

Participating in an alliance is often a good idea. The following list reflects what the buying coalitions and the payors are really looking for in future subacute care partners:

- Single source to the extent possible
- Standardization of quality
- Cost efficiency
- Consistency of process and procedure

What does participating in an alliance with an acute care hospital really do for a subacute care provider? The advantages can be summed up as follows:

- Accelerates access to market to deliver lower cost post-acute services and programs

- Provides an opportunity to target program design to meet the service expansion needs of the local or regional network
- Enhances local and regional market exposure
- Provides a reasonable basis for risk sharing in the full continuum of care delivered
- Offers the potential to significantly change the census mix

Payors are really looking for certain relationship features that a subacute care provider should be attentive to. These are the seven smart principles:

1. *Know me.* Know not only the organization's needs, but also the strengths and the weaknesses. The payor wants to make certain that the provider understands the payor's management and how decisions are made.

2. *Tell me what you can do for me.* In order to satisfy this need, the provider must be specific and have outcome information readily available, according to the payor's general parameters and specifically by locations.

3. *Relate to my management.* The provider must be able to not only know the management, but to know the management's vision, mission, and objectives.

4. *Communicate on the payor's terms,* to whom, how often, and in what form.

5. *Respond to requests even though the request may seem unreasonable.* The provider must understand the payor's motives in seeking the information. Talk through the need and then agree on the best method of addressing it.

6. *Get the payor good information and on time.* This information is needed to reinforce the original decision for participation in the relationship. This information, as stated above, should involve both location and the ability to be payor specific.

7 *Stay in touch.* This means keeping the communication lines open in good times as well as bad. Constant communication will allow the payor and the provider to service concerns before they arise, and to discuss the available solutions so that there are as few surprises as possible at the point of service.

SUMMARY

Subacute care works well in this new era of cost-conscious healthcare. What is even more important is that the key issues and critical success factors for subacute care providers are the exact elements that will make the provider an attractive partner to the payor. When examined closely, the goals and objectives of subacute care providers and payors are closely matched.

Subacute care programs have gained the respect of insurers, HMOs, and PPOs as a proactive approach to delivering services to enrollees. Some of the largest payor companies in the United States have acknowledged the benefits of subacute care services in their contractual language. MeTra Health, formerly MetLife, has even established a "think tank" to evaluate and research the impact that subacute care will have on its organization.

Subacute care has developed as a result of the continually rising costs of healthcare and managed care companies' demand for appropriate programs to be provided in a cost-effective setting. Subacute care has accomplished this efficiency without compromising the quality of care the patient receives.

CHAPTER 8

Legal Considerations

"I know the difference between good and evil."

— Dr. Who

Subacute care is an integral part of integrated delivery systems that are receiving increased attention from managed care payors and other referral sources. This chapter will address legal issues regarding integrated delivery systems and managed care contracts.

It is important to note, however, that this is not intended to serve as legal advice. Subacute care providers and integrated delivery system managers should seek legal counsel regarding legal matters.

Depending on legal constraints, the integrated delivery system may create a vehicle for some providers in the system to direct capital for the benefit of the whole system. Additionally, sufficient financial integration among providers in the system may reduce the risk of antitrust violations that stem from price-fixing. Finally, owners of a proprietary integrated delivery system model may build value in the integrated delivery system entity for purposes of a future sale or negotiation with another healthcare organization.

The integrated delivery system often includes (consistent with legal constraints) incentives to solidify providers' use of the system for management and managed care contracting, and for appropriately keeping patients within the system.

All of this will enhance the ability of the integrated delivery system to effectively compete for managed care contracts, especially at-risk financial arrangements. Moreover, most analysts agree that those provider organizations that can effectively compete for at-risk financial arrangements will be well positioned for any future government healthcare reform.

STRUCTURAL MODELS

For the purpose of this discussion, integrated delivery systems refer to legal entities such as management services organizations (MSOs) and foundations. Several structural models have been developed for the organization of integrated delivery systems. This discussion will be limited to two of the most prevalent models: MSOs and medical foundations. Most other market models—physician-hospital organizations (PHOs), primary care-based networks, regional delivery systems, regional networks, and so on—usually include one or more of the features of MSOs or foundations.

Management Services Organizations

An MSO is generally described as a new legal entity formed for the purpose of jointly marketing and managing providers. MSOs usually include at least one hospital and one professional medical organization. The MSO can be organized

EXHBIT 8.1
MANAGEMENT SERVICES ORGANIZATION/SERVICE BUREAU

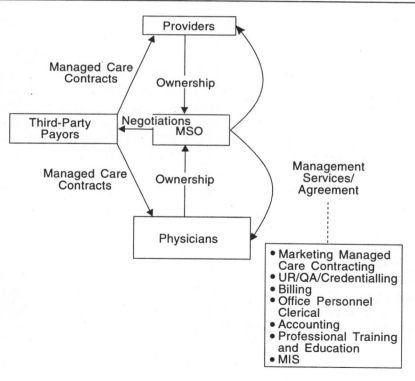

EXHIBIT 8.2
MANAGEMENT SERVICES ORGANIZATION/ASSET ACQUISITION

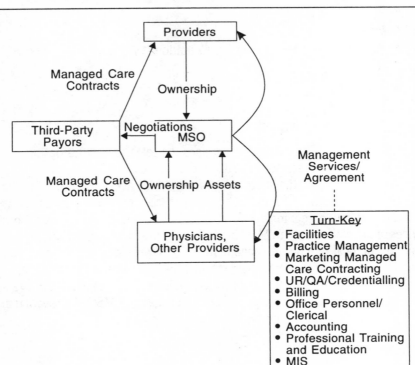

as a for-profit corporation, nonprofit corporation, or general or limited partnership. Some or all of the providers may be shareholders, members, or partners of the MSO, but all of the provider participants retain their status as separate organizational entities. The MSO typically is not a provider of healthcare services: All contracts with health plans and other payors, although negotiated by the MSO, are entered into directly between the MSO's providers and the payors.

There are two basic types of MSOs. The first type can be termed a *service bureau MSO.* It furnishes at least marketing and managed care contracting services (including development of the provider network for the MSO) under power of attorney, as well as coordinating utilization review, quality assurance, and provider credentialing services to the providers in the MSO, pursuant to a management services agreement entered into between the MSO and each provider. The MSO also may furnish other management and administrative services to its providers, such as:

- Billing
- Claims processing and collection
- Personnel
- Equipment
- Computer and management information systems
- Accounting
- Legal services
- Insurance
- Miscellaneous vendor services

Some MSOs offer a menu of services for selection at the option of the providers. This type of service bureau MSO is depicted in Exhibit 8.1.

The second type, the *asset acquisition MSO*, acquires various assets of some or all of its providers, including its physician providers, under an asset purchase and sale agreement. This type of MSO furnishes most or all of the services of the service bureau MSO but also furnishes management services—for example, medical practice management—to the providers whose assets have been purchased, under a management agreement between the MSO and the providers. This asset acquisition MSO generally is more capital-intensive and furnishes more extensive management services than the MSO. The second type of MSO is illustrated in Exhibit 8.2.

Foundations

A foundation, like an MSO, is a new entity formed to jointly own and manage a group of providers. Many foundations are organized as nonprofit, tax-exempt entities. Typically, the foundation purchases the assets of its physician organizations and possibly other provider organizations as well. Unlike the MSO, the foundation itself becomes the provider of services and enters into contracts directly with third-party payors. Upon acquisition of the provider assets, it employs all nonphysician and perhaps other provider personnel, operates the medical practice or other providers' businesses, and typically enters into an agreement for the physician organization to provide professional services to the foundation in exchange for compensation. The physician organization in turn employs individual physicians to perform professional services.

A foundation seeking tax-exempt status may need to meet strict criteria for exemption according to the Internal Revenue Service's recent exemption determinations for foundations.

The foundation is more capital intensive than either the service bureau MSO or an asset acquisition MSO. It is more fully integrated and therefore can be more effective in reducing costs, operating in a more coordinated and efficient

EXHIBIT 8.3
FOUNDATION

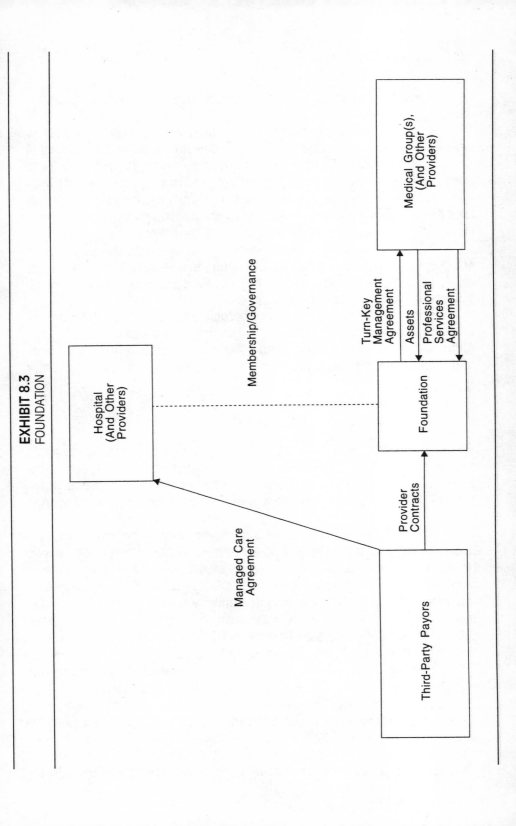

manner, and competing for managed care contracts. The foundation model is illustrated in Exhibit 8.3.

Subacute care providers may align themselves with integrated delivery systems in several ways. One way is to become an owner and a contracting provider of the system. This may be a preferred alternative to the degree it can position the subacute provider to participate in system profits derived from operations and future sale, and to access monies from risk funds and risk pools established from the provision of services by the system's providers. Ownership also usually means a voice in the governance of the system.

Second, the subacute care provider could become a contracting provider of the system, without ownership. This option may involve less initial capital outlay for the subacute care provider but likely will preclude some or all of the benefits of ownership just discussed.

Third, the subacute provider can become a contracting provider of subacute care for one of the providers in the system, for example, a hospital. Although the latter will produce even fewer of the benefits described in either of the other two options, it can provide an entrée into an integrated delivery system.

THE ISSUES OF ANTITRUST

The antitrust laws, both state and federal, generally prohibit unreasonable restraints on trade or competition. Certain conduct—price-fixing, for example—will be unlawful per se irrespective of competitive or anticompetitive effects. Other conduct requires review of various factors under a "rule of reason" analysis.

Integrated delivery systems present certain areas of potential exposure under the antitrust laws. The collective discussion of price-fixing to be charged by the providers in an integrated delivery system to third-party payors may violate the law. Integrated delivery systems that are sufficiently financially integrated—for example, those operated under a capitated payment arrangement—may be treated as a single economic unit, incapable of conspiring to set prices. Insufficiently integrated systems may be subject to price-fixing allegations. To comply with price-fixing restrictions, most integrated delivery systems utilize messenger or modified messenger models to obtain price information and establish rates for purposes of negotiating with managed care payors.

Large provider networks that compose integrated delivery systems may also be seen as monopolizing the market and excluding competitors unlawfully under the antitrust laws. The key issue here is whether the system has sufficient market power: the ability to set prices without affecting output. These prohibitions may restrict provider credentialing and affect the integrated delivery system's actions to include or exclude certain providers in the system. Similarly, the integrated delivery system may be subject to an allegation of "group boycott."

To reduce legal risk in this area, integrated delivery systems should carefully review whether they have sufficient market power, adopt objectives and measurable criteria for credentialing purposes, and should confer decision-making authority on persons or organizations that are not competitors in the marketplace.

TAX

Some integrated delivery systems such as foundations may seek to obtain tax-exempt status. The Internal Revenue Service (IRS) has made recent exempt determinations for medical foundations using the following factors, among others, in determining whether the foundations under review should be granted tax-exempt status:

- ◆ Whether the integrated delivery system participates in Medicare and Medicaid programs
- ◆ Whether emergency care is furnished irrespective of ability to pay
- ◆ Whether the system has adopted a charity care policy
- ◆ Whether the system's acquisition of physician assets is at or below fair market rates
- ◆ If physician compensation is at competitive rates
- ◆ Are no more than 20% of the physicians participating on the governing board of the integrated delivery system

It is likely that the IRS may use these and other similar factors when reviewing tax-exempt applications from integrated systems in the future.

Other rules apply to tax-exempt integrated delivery systems seeking to obtain tax-exempt financing. In addition to not having more than 20% physician representation on the governing body, Revenue Procedure 93–19 establishes further limits on the length of agreements with physicians and other private organizations as a condition of tax-exempt financing.

When an integrated delivery system—for example, a foundation—is granted tax-exempt status, all expenditures of the system must be for fair market value. Otherwise, the system will violate federal tax prohibitions on inurement and private benefit, thereby jeopardizing the system's tax-exempt status.

Integrated delivery systems that are tax-exempt may be subject to possible exposure. For example, the purchase price for physician assets should be at fair market value and determined by independent appraisal. Income guarantees similarly may raise issues of inurement and private benefit. Compensation from or to physicians and other providers under the management agreements and professional services agreements should also be at fair market rates. Phantom capitalization by physician organizations and other providers in the integrated delivery system will create risk. Additionally, other compensation and financial relationships with physicians and other providers—for example, medical directorships and office leases—should also be at fair market rates.

Management services organizations will find it difficult to obtain tax-exempt status because they are commonly viewed as servicing such nonexempt organizations as physician organizations.

Management services organizations should not present a great risk in the area of corporate medical practice. However, to the extent that MSOs interfere with the clinical aspect of the medical practice, as in the performance and utilization of quality assurance, corporate practice issues may arise. These issues also arise in MSO models where the integrated delivery system or other lay organization effectively purchases the revenue streams derived from the performance of professional services.

Foundations may present a greater risk of violating corporate practice doctrine to the extent that they provide medical care, but in states like California, some experts have concluded that there is an implicit exception to the application of the corporate practice doctrine for foundations organized under certain provisions of state law. Many integrated delivery systems attempt to restrict physicians who participate in the system from participating in other delivery systems or from competing with the integrated delivery system. These restrictions, if they are construed as interfering with the physicians' ability to practice medicine in other settings, also may raise corporate practice of medicine issues.

ERISA

The Employee Retirement Income Security Act of 1974 (ERISA) is a federal statute that regulates employee benefit plans, including retirement plans and employer-sponsored health plans. The formation of integrated delivery systems, particularly MSOs, may create one or more "affiliated service groups" under the terms of the law that could adversely affect the favorable tax treatment accorded the pension and profit sharing plans of each of the providers participating in the integrated delivery system.

If an affiliated service group is created, certain nondiscrimination rules concerning benefits, coverage, and contributions must be met respecting all employees of providers within the affiliated service group. Compliance with these rules may be an expensive and extremely burdensome process. Providers, therefore, are well-advised to carefully review the organization of an integrated delivery system in view of requirements under ERISA and the Internal Revenue Code.

MEDICARE REIMBURSEMENT

Medicare reimbursement rules present significant complications for integrated delivery systems. MSOs likely will be unable to obtain a separate provider number for purposes of billing services furnished under the MSO. Rather, the services of each participating provider will be billed under its own provider number. Medicare's limitations on assignment of Medicare claims to the integrated delivery systems exacerbates this problem. Further, Medicare's "incident to" rules will require that services of certain allied health professionals, such as those of physician assistants, must be performed by the physician organization's own employees, not personnel leased or contracted from the integrated delivery system.

Without legislative or regulatory changes, integrated delivery systems may have a difficult time functioning as a truly integrated, unified entity for purposes of Medicare reimbursement.

GENERAL ISSUES

Other legal issues that arise when organizing integrated delivery systems. State laws regulating healthcare plans, insurance, and/or licensure may limit the system's ability to coordinate the delivery of care from multiple providers. These laws may establish financial viability standards for the system and its providers, or may prevent the system from assuming financial risk or global fee services furnished by its providers.

Exclusivity and noncompetition clauses in an integrated delivery system's management agreements with its providers may invoke concerns regarding violation of the corporate practice of medicine doctrine and fraud and abuse laws, and may have questionable validity under state law.

Although the federal certificate of need statute was repealed in 1986, many states have retained certificate of need laws. Integrated delivery systems should be cognizant of applicable state laws as they may relate to the organization, expansion, and contribution of service categories and facilities within an integrated delivery system.

MANAGED CARE CONTRACTS: TERMS AND CONDITIONS

Subacute care providers seeking to contract directly with managed care payors or through an arrangement with an integrated delivery system should carefully consider the terms and conditions of those agreements during the negotiation process, before executing the agreement. Summarized here are examples of some of the key contract provisions which subacute care providers should consider including in their agreement with managed care payors.

Corporate Status

Each party should identify its corporate status in the contract. Nonprofit providers should indicate they are nonprofit. Each party should be required to notify the other of a change in corporate status. PPO contracts should be executed with the party responsible for making payment.

Licensure

Each party should indicate how and under what regulatory authority it is licensed, if at all. This will help enable the subacute care provider (known in contract terms as the SCP) to identify the laws with which the managed care payor (Payor) is obligated to comply. The SCP should be entitled to receive proof of any such licensure or regulatory status from the Payor, and the Payor should be required to notify the SCP of a change in its licensure or regulatory status.

Access to Payor's Financial Statements

The SCP should attempt to access the Payor's relevant financial statements in order to determine its financial viability and financial "track record."

Definitions

The contract should include a section that defines all relevant terms or phrases occurring in the contract, for example:

- Medical necessity
- Emergency
- Member/enrollee
- Covered stay

Covered Services/Level of Care

The contract should specifically enumerate all services and the level of care which the SCP is responsible for providing under the contract, for example:

- Ordinary nursing
- Extraordinary nursing
- Occupational therapy
- Physical therapy
- Other ancillaries
- Supplies

This should also encompass services that, if normally provided by a hospital, are excluded from coverage.

Third-Party/Subscriber Agreements

If coverage will be determined according to the terms of third-party/subscriber plans, the contract should permit the SCP to obtain copies of such contracts or plans for its evaluation. Preferably, such plans should be incorporated by reference in the contract. No new services/benefits or classes of members/enrollees should be added without the SCP's prior approval.

Limitations on Coverage

The SCP should determine whether it is willing and able to assume responsibility for furnishing covered services that are not normally provided by the SCP. If not, the scope of coverage should be limited accordingly. (See the discussion later in the chapter about the transfer or delegation of services.)

Guarantees

The contract should clarify whether the SCP has the discretion to refuse to admit or accept for treatment members/enrollees who are otherwise eligible to receive services under their respective arrangements with the health maintenance organization (HMO) or preferred provider organization (PPO). In any event, the contract should provide that the SCP may refuse to provide care to HMO/PPO patients if hospital beds, facilities, or services are not available.

Specialty Services

The SCP should ensure that the contract clearly indicates whether the SCP is responsible for providing other ancillary services. This is especially important in a capitation arrangement. The SCP should ensure that its Payor contractual obligations are consistent with those obligations owed to physicians.

Religious and Ethical Considerations

If applicable, the contract should provide that the SCP is not required to provide any services inconsistent with the religious or ethical principles of the organization.

Referrals/Transfers

Patient admission, referral, and transfer policies/responsibilities should be specified in the contract and should be consistent with the SCP's policies and procedures. The SCP may need to develop a separate contract with the patient regarding compensation issues and ensure consistency with bedholds. The SCP should retain the authority to discharge/transfer the patient in the event SCP services are no longer covered or to protect health interests of other residents.

Arrangements for Covered Services

The SCP should be specifically permitted to transfer patients and/or delegate-covered services to other providers when the SCP cannot provide (or is unwilling to provide) such services. The contract should identify which party is responsible for arranging such transfers and delegation, and whether the transferring SCP or receiving provider is ultimately responsible for providing the service, as well as which facility shall receive payment. The SCP should receive payments from the Payor for any services rendered pending such transfers.

Arrangements for Noncovered Services

Similarly, the contract should clearly state that any noncovered services provided to the Payor for patients will be paid pending transfer.

Payment

The SCP should identify the proposed payment rate. Examples include

- ◆ Discount from charges
- ◆ Per diem rate
- ◆ Capitation
- ◆ Case rate

The SCP should then carefully analyze the payment rate, in conjunction with anticipated utilization patterns, to determine whether the proposal is financially desirable. The SCP should take particular note of any capitation or other at-risk arrangements. An actuarial analysis, examining—among other things—utilization patterns, care-mix, and percentage of premium retention, should be undertaken.

Capitation arrangements may result in the inclusion of additional provisions, such as withholds/incentives. The SCP should protect against physician/professional overutilization in these circumstances. Also, the contract should establish clear requirements and limitations for the administration of any withhold account.

Acceptance of capitation from "nonlicensed" HMOs or any payment from capitated, nonlicensed providers may raise issues under state HMOs, insurance, and/or licensing laws. Administration of a risk pool also may pose risks under these laws.

Billing, Payment, and Collection Procedures

The agreement should clearly describe the billing, payment, and collection procedures and the parties' respective responsibilities. Preferably, the SCP's existing billing form should be used to report discharge summaries and other such pertinent billing information to the Payor. Billing should be submitted promptly to the Payor—such as by the end of the month in which discharge occurs or within a certain number of days after discharge. The SCP should be permitted to report on long-term patients at regular intervals during the patient's stay.

It is important that the contract obligate the Payor to make payment promptly (within 30 days, but not later than 45 days after receipt of bill). The SCP may wish to require the Payor to pay interest on delinquent accounts. The SCP should have the authority, to the extent permitted by law, to bill the patient if the Payor fails to make payment. The contract should authorize the SCP to take assignment of the Payor's patients' benefits and to bill such patients for applicable copayments. The contract should also specify who is responsible for undertaking collections of any unpaid copayments; if the SCP is responsible, the Payor prefer-ably should bear responsibility for collection costs. The contract should permit the SCP to bill the patient directly for noncovered and medically unnecessary services. The SCP should be permitted to bill the patient after advance written notice to the patient. There should be no requirement that the SCP obtain the patient's prior consent before billing.

Assurance of Payment
It is important to ensure the Payor's continued financial viability under the contract. The SCP probably will be unable to obtain full payment for prior services rendered in the event of bankruptcy. Contract provisions permitting the SCP to terminate in the event of insolvency may be unenforceable. The SCP may wish to require the Payor to post a performance bond, guaranty, or deposit, especially in the event of potential insolvency of the Payor. The Payor should be required to pay all or some (such as 80%) of the rate in the event of a utilization review dispute. Stop-loss or reinsurance protection is especially important in per diem rates, such as those for a pharmacy or capitation arrangement.

Insurance
The SCP should ensure that the Payor meets the following requirements:

- The Payor shall maintain comprehensive general liability and profes-sional liability coverage.
- The liability limits of such insurance policies are subject to the SCP's approval.
- The Payor should provide the SCP evidence of such coverage and limit upon request, in a form determined by the SCP
- The Payor should notify the SCP immediately upon any significant changes in insurance, such as cancellation.

Preferably, the contract, especially those involving capitated arrangements, should provide for stop-loss or catastrophic insurance coverage. Typically, such coverage will provide for the contract rate to revert to billed or a percentage of billed charges for the entire stay or from the time the excess level is met, in the event usual and customary charges exceed a certain level, such as $100,000, for any individual patient, or if costs per day exceed contract's per diem rate, e.g., $500. Or the contract may provide for a more similar reversion to a more favorable rate in the event the difference between aggregate contract rate charges and billed charges exceeds an agreed upon amount.

Indemnification

The SCP may attempt to ensure that the HMO/PPO agrees to indemnify the SCP for any of its liabilities or losses stemming from the acts or omissions of the HMO/PPO, or its contractors. The SCP should preferably refrain from agreeing to indemnify the HMO; if it does, it should verify that such indemnity is consistent with its insurance coverage.

Coordination of Benefits

The contract should specify how benefits from primary payment sources will be coordinated with payments made by the Payor to the SCP. The SCP should obtain answers to the questions such as: Will the Payor make full payment at the contract rate and receive a "credit" for such primary payments later received by the SCP, or will the Payor be entitled to reduce its payment to the SCP in the amount of any primary payments previously made to the SCP? If the contract rate is less than the SCP's actual charges, will primary payments be credited to the SCP's account for the difference between the contract rate and actual charges?

Preferably, the SCP should be permitted to collect from the Payor if third-party payment is available, and the Payor should be ultimately responsible for collections.

Grievance Procedure

The contract should establish and describe a dispute resolution or grievance procedure for resolving disputes arising between the parties under the contract. The contract should initially establish and describe an informal grievance mechanism, including but not limited to a description of the following:

+ Each party's liaison person
+ Composition of the body or decision-making entity that will resolve disputes

- Identification of who may initiate disputes and how they may be initiated
- Identification of the time limits for processing grievances

The SCP also should ensure that the grievance mechanism is compatible with similar processes already established by the SCP, the Payor, and insurers/payors.

Disputes not satisfactorily resolved through the informal grievance mechanism may be referred to arbitration organizations such as the American Arbitration Association (AAA), or to mediation. Arbitration costs should be shared by the parties, except that the grieving party should pay fees initially, subject to reimbursement by the other party upon termination of the AAA proceedings. The SCP should attempt to require representation on the arbitration panel by those with expertise. Binding arbitration has its advantages. Sometimes it is more expedient and less expensive. Malpractice disputes should be subject to the applicable party's own patient arbitration agreements, unless those agreements are inconsistent with the law or the other party's arbitration program.

Medical Records and Confidentiality

The SCP should assume responsibility for maintaining patient medical records. The contract should state that medical records are the SCP's property, but that the Payor's enrollee records may be inspected by the Payor pursuant to the contract, consistent with applicable state law, and in a reasonable manner.

The contract also should provide that the SCP will disclose information contained in the Payor enrollees' medical records only if such disclosure is consistent with the law. The Payor should be responsible for obtaining any patient releases that may be necessary for disclosure and should not be permitted to disclose medical information to third parties without prior approval from the SCP. At the least, any such disclosure should be made consistent with the law, and the SCP should be notified promptly of any such disclosures. The provider should be permitted to disclose such medical information where disclosure is consistent with law. Finally the Payor should not be permitted to review or obtain copies of the records or proceedings of professional staff committees.

Access to Records

The contract should provide that audits and review of all of the SCP's records and professional staff committee records, as previously discussed, should be made by the Payor, consistent with the considerations of confidentiality, in a reasonable manner within the proper scope of the Payor's review authority and consistent with all other applicable law. Audits and all records review should be

- Performed at reasonable times only
- Performed only after prior notice to the SCP
- Limited to the Payor's authorized scope of inquiry
- Performed at the Payor's expense, such as photocopying

Information obtained by the Payor through audit or review should not be released to third parties absent the SCP's prior consent.

Utilization Review

The Payor's proposed utilization review (UR) plan should be fully described in the contract or in an attached exhibit. The UR plan should discuss whether review is delegated to the SCP, the methods of review (preadmission and/or concurrent review), and the situations in which such different forms of review will apply. The SCP should avoid retrospective review. The standards of review should be defined. The contract also should specify the qualifications of review personnel and whether review will be conducted on site. If review will be on site, the contract should also limit the Payor's review of information in physicians' offices and records.

The Payor should bear photocopying and other costs incurred in complying with the UR plan. The contract should integrate the existing SCP's UR procedures with the Payor's UR plan. Further, the contract should address patients' rights in the UR process.

Other Contract Considerations

The contract should clearly delineate an effective appeals process to contest Payor denials and specify how the SCP may change or challenge standards, such as for diagnoses requiring preadmission certification. It should provide for confidentiality of the peer review process and protection from third-party access to Payor records. The notice of denial to the patient should come from Payor. The contract also should require professional staff to comply with the SCP's utilization review policies and procedures.

Patient Identification/Eligibility

The Payor should notify the SCP of all Payor patient members/enrollees and any changes. The means of notification and verification of eligibility, such as a membership card or telephone call to the Payor, should be specifically set forth in the agreement. Verification should be done promptly. The SCP should be guaranteed payment for services provided to patients who are identified or verified by the Payor as members/enrollees (even if information provided by the Payor may be erroneous) until notification of noncoverage is provided by the Payor or alternative coverage has been arranged, or a medically acceptable transfer or other disposition of the patient has been made. Members/enrollees who are later found to be eligible should be covered; that is, the Payor should not be permitted to disenroll its members/enrollees retroactively. Emergency services should be paid based on the medical opinion of the physician, not the Payor.

The contract's payment rate should not apply where services are provided to a patient who was in the SCP program at the time the contract is executed, unless specifically identified such. If the patient's membership/enrollment status terminates during the subacute care stay, the contract rate should not apply with respect to services performed after the termination of enrollment status. Non-solicitation provisions can be an effective way to protect the SCP's patient base.

Professional Staff
The contract should specify that all SCPs, physicians, and other allied health professionals must comply with professional staff requirements and other policies and procedures that the SCP states as preconditions to performing any services at the subacute care facility.

Advertising
The HMO/PPO should be required to provide the SCP with all advertising and marketing materials disseminated to the public, potential Payors, or patients. Any such materials containing references to the SCP should be subject to the SCP's approval prior to dissemination.

Exclusivity
The parties may intend that covered services will be provided (in a prescribed geographic area) only by the specific SCP or by providers participating in the SCP's integrated delivery system. Such arrangements may raise antitrust and fraud and abuse concerns. Therefore, counsel may be necessary to discuss the propriety of any such provision and to draft appropriate language.

Term Renewal and Termination
The contract should specify an initial term, preferably one or two years. There are pros and cons to automatic renewal clauses, so they should be given careful consideration.

The contract should be terminable in the event of material breach by either party, for example: "Termination will be effective upon 30 days notice unless the other party cures the breach within 20 days after receipt of notice to terminate." The payor's default in payment can trigger termination that, for example, may be effective upon five days notice if payment has been delinquent for the prescribed period of time. If the contract violates the law, the contract may terminate. Either party should be permitted to terminate without cause with sufficient advance notice, such as 45, 60, or 90 days. Shorter notice periods may be appropriate for the Payors with questionable reputations or "track records." The contract should provide that the Payor immediately notify the SCP in the event of insolvency or pending insolvency (as defined in the contract), and the SCP should be permitted to terminate the agreement immediately in the event of such insolvency. This may permit the SCP to bill members/enrollees directly, which may be otherwise prohibited during the term of the contract by its own terms.

Upon termination, the contract should cease to apply except as follows: The contract should state that each party will remain responsible for the obligations or liabilities arising from its or its employees' servants, or agents' actions or inactions occurring prior to termination. The Payor should remain responsible to make payment at the contract rate for services provided by the SCP subsequent to termination and until suitable transfer can be arranged. The contract should clarify that a member's/enrollee's physician may have access to medical records within a prescribed period after termination of the contract.

SUMMARY

Whether a subacute care provider is able to successfully negotiate those provisions will depend on the relative bargaining strength of the parties. Subacute care providers that are part of larger networks or integrated delivery systems may increase the likelihood of achieving their contracting objectives.

CHAPTER 9

Marketing

"It would spoil my day if I missed this one."

— Doc Holiday

The development of a strategic marketing program for subacute organizations is a unique undertaking that requires both energy and resources. The program itself evolves over time as individual aspects of marketing are developed. In some areas subacute care is in a nascent stage of development; in other areas it is becoming more defined.

In terms of the product life cycle, subacute care is moving from the introduction stage into the rapid growth stage. Many of the efforts involved will be to create awareness within the marketplace about the utility and benefits of the service. This includes both cost containment elements and a more effective environment for the patient at that particular level of recovery.

The ability of organizations within the subacute care industry to develop marketing programs is driven in part by experience. Most marketing activity in the past was more internal and involved taking referrals over the phone and processing an admission. Not that long ago the demand for beds was greater than the supply, and the facility providers were the beneficiaries.

MARKETING COMMITMENT

An organization's commitment to marketing begins at the highest levels of the organization. Adequate resources must be devoted to both the internal and external aspects of the marketing program. Senior management must be cognizant that this is a long-term investment that requires time and financial resources.

Marketing itself is a process that contains a series of functions and multilayered activities. It is important to point out that there are many aspects to the development of a marketing program. Direct marketing is only one component of a comprehensive program. Other aspects of a full range of marketing services include the research function, promotion, and public relations. Marketing services often include the development of brochures and collateral materials, internal marketing, planning, information systems, strategic plans, and a product line or brand management.

Subacute care programs need marketing to serve and encourage referrals from healthcare professionals, families, insurance case managers, physicians, and the community. In cultivating awareness, marketing professionals should represent a service that captures a preferred status.

Marketing is a function that requires timing, skill, and experience to be successful. Flexibility, especially as it relates to customers, is critical. A combination of resources can be arranged in myriad ways and are usually facility or program specific.

The following are some factors that have proven to be successful for subacute care marketing efforts:

- Commitment to marketing from top management
- Substantial financial investment in physical plant and personnel
- Ability to screen patients quickly and respond to critical pathways
- Collateral materials aimed at targeting the appropriate audience
- Clinical inservices with referral sources
- Encourage facility tours
- Identifying and promoting the unique attributes

The value system of the organization becomes a critical component of success. Marketing values should be instilled within the organization and help improve efforts to be market driven. This includes embracing change and reconstructing services to meet new needs.

Values related to achieving census goals and revenue projections should be encouraged. These should frame the attitude of the marketing department as well as the attitude of the clinical caregivers who are actually providing the treatment.

The ability to overcome obstacles and barriers is an important marketing skill. The ability to improve or change referral patterns often requires creativity. The marketer must be willing to make the fifth, sixth, and seventh attempts to achieve success. Barriers should be viewed as opportunities to educate referral sources, including physicians, hospital social workers, and managed care case managers.

Customers should be treated with concern, understanding, and respect. Development of relationships with customers occurs over time and is built on trust. There are always new markets to penetrate and new customers to cultivate, as well as existing customers from whom to acquire more referrals. In fact, it is much easier to increase the level of referrals from existing customers than it is to establish new relationships.

It is important to be flexible in order to complete the referral process. This can take the form of paying attention to the smallest detail or of persuading the president of an organization to choose subacute care. The management of customer relationships is a global process and one that is built one plank at a time. The marketer should make decisions based on evaluating the risk in terms of the costs and the benefits for the organization. It is often difficult but can yield rewards and result in a long term relationship with the customer.

The marketing person functions as a facilitator and should encourage frequent interactions between the customer and the subacute care staff. Critical competencies of the organization can then be identified and create a competitive advantage for the organization.

Organizations have an opportunity to achieve customer satisfaction in all areas. Communication with customers is critical to an ongoing, long-lasting, and mutually beneficial relationship. Customers will invest as much time and effort in the relationship as the subacute care program demands. People who hold decision-making positions within the organization and who recognize the importance of these relationships will succeed in improving the environment within their organization and meet their marketing goals.

DEVELOPING A MARKETING PLAN

Marketing cannot operate in a vacuum. The other departments such as nursing, systems information, and ancillary services play a significant role in developing the marketing plan. Financial goals and the budget forecasts for census and revenue should also be included as part of the process. The marketing plan should assist in realizing the potential within the service area to increase revenue. Additionally, the marketing department is in a good position to identify service delivery gaps that can be altered.

The following activities may prove helpful when planners develop a marketing plan:

- The plan should decrease time frames for goals and objectives that will improve predictive abilities.

- Create a strategy promoting long-term relationships with referral sources. This will create value and secure the program's future.

- Include other departments in the development of the marketing plan.
- Constantly seek ways to meet the service needs of referral sources. The creation of value hinges on enhanced communication.
- Marketing goals must be supported with adequate financial resources. Develop budgets to support the direct and indirect marketing efforts.

The marketing plan discussed here has a one-year planning horizon. There are tactical plans that address more specific targeted goals identified on a 30- or 90-day time frame. It is important to address all components of the marketing mix within the planning horizon.

The role of marketing and the various functions within a comprehensive marketing program are usually unknown to other departments. They serve to identify how broad the true scope of marketing should be within the organization. These functions include

- Market research
- New program development
- Developing collateral materials
- Advertising
- Promotion
- Information systems
- Budget and planning
- Forecasting
- Managing product lines
- Designing pricing strategies
- Analyzing consumer behavior

Vision, Mission, and Goals

The first three segments of a marketing plan are the vision, mission, and goals. *Vision* encompasses the organization's size, scope, desired future, and so on. The *mission* describes what the organization is and serves as a framework to evaluate future opportunities. A mission should also describe how the organization views itself in relation to its position in the marketplace.

The *goals* depend on the planning time frame. Long-term goals may help to signal a shift into new markets or services. They may describe a pricing posture or an image. Short-term goals are more tangible and time specific. They normally describe census goals, projections, case mix changes, or other key indicators. They may also relate to marketing image or geographic exposure. Goals should serve as the "ends," whereas other aspects of the marketing plan should focus on the "means" to achieve these ends.

Market Service Area

Normally, this segment of the marketing plan relies heavily on past data. Identification of the catchment area from a review of past referrals and admissions is helpful. This process allows identification of the key referral sources. Other useful data include evaluating payor mix and diagnosis related group (DRG) information, identifying physician specialists practicing within the catchment area, and compiling census data and mailing lists.

Market Share Analysis

In most situations limited data exist regarding patient volume of specific categories by payor type. More pronounced information is available in terms of DRG classification. It's extremely useful to identify the potential market share of the subacute care setting. Opportunities exist to estimate this by focusing on the other subacute care programs in the catchment area in terms of occupancy and payor mix. Given existing classifications and mixed models, it becomes challenging to estimate. Market share is important because it gives an indication of size, scope, and market position to the marketing plan. Short and long term goals may also be stated in terms of relevant market share and payor mix goals.

SWOT Analysis

SWOT stands for strengths, weaknesses, opportunities, and threats. The market planning process presents a unique opportunity to evaluate the facilitiy's core competencies in relation to its competition. Keep in mind that the competition can take various forms:

- Competing providers within the industry
- Available substitutes for subacute care services
- Payor mix competition
- Competition by category of illness or injury
- Short or long term care programs
- Specialty subacute care programs versus general subacute care programs
- Hospital-based subacute care settings versus subacute care programs within freestanding skilled nursing facilities

In terms of strengths the marketing plan should focus on core competencies such as:

- Staff skill set
- Physical plant advantages
- Service dimensions

- Referral process
- User-friendly access
- Experience of subacute care staff
- Length of time the subacute care program has been operating

Careful consideration should be given to those items that the customer sees and evaluates. Weaknesses should be identified and subsequently changed or improved. Once they are isolated and identified, a time frame for change or improvement should be established.

The method that is adopted in order to address the problem should be accepted and employed by the interdisciplinary team. Most problems cross departmental lines and require attention from the subacute care staff as a whole.

Opportunities should be considered in light of organizational resources and gaps in services within a catchment area. Research into new service development should consider the existing reimbursement environment in terms of third-party payors and Medicare. In addition, consideration should be given to the regulatory environment, to changes in health department rules, licensure, or certificate of need. Other factors such as demographic or socialization trends figure in new service development.

The competition is most always vulnerable in some aspect of programming. Increasing by decreasing is a legitimate strategy. Penetration in the soft spots of strong competitors is a successful strategy to win market share. In addition, new uses for subacute care services must also be considered. An example might be direct admission to a subacute care setting for a procedure or recovery period that in the past was treated solely in the acute setting.

Providers should attend to areas that could reduce market share in a catchment area. Opportunities may arise that complement or add new services to programs. These changes represent the ebb and flow in the balance of power within catchment areas. Providers should always have some contingency plans prepared.

Forecast

The marketing program should work closely with finance and accounting to establish reasonable goals. These goals can be stated in the marketing plan in terms of revenue, margins of profitability, payor case mix (insurance/medicare), type of patient, source of revenue (routine versus ancillary revenue categories), and type of payment and/or risk (capitation).

Once goals are agreed upon they should be shared with the marketing staff. These goals serve to identify the criteria for establishing components of the marketing plan. The staff should agree that the revenue goals are reasonable and achievable.

Marketing Budgets

Any revenue goal requires a certain degree of support in terms of financial resources devoted to the marketing efforts. The marketing budget that becomes part of the marketing plan process should reflect the forecast. Normal marketing budgets can be expressed as a percentage of revenue or as a specific amount per month. Another approach is to develop budgets either based on the project or category of marketing expense such as public relations or sales promotional items. As with most healthcare departments, the major portion of most subacute care marketing budgets are direct salaries or expenses related to employing marketing personnel and marketing support staff. The marketing budgets should include allocation for salaries, travel, public relations, consultants, research, advertising, collateral materials, conferences and exhibitions, and memberships in outside organizations to increase visibility. There is a plethora of other, smaller expenses that are a part of any comprehensive marketing effort.

Competitor Analysis

This area of the plan focuses attention on the competition. The key competitors should be identified and all relevant data about the competition should be discussed. Relevant data include results of a visit to the competitors' physical locations, evaluation of key competencies and reputation, market share, and so on. This review will assist the marketing staff to develop tactical plans and strategies.

Strategic and Focused Planning

In this component of the marketing plan, short-term (90 days or less) tactical plans are developed. These plans include summarizing key events and customer activities, conferences, focusing on monthly events, customer satisfaction surveys, key account plans, direct mail campaigns, advertising and public relations, and the increased generation of qualified leads. These efforts require ongoing support and coordination to be effective. Limiting the number of activities allows the provider to maximize the benefits of each activity chosen.

A Full Census

This section of the plan requires the ability to calculate the ratio of patient referrals that convert to admissions per year and the number of admissions that are required to keep the program at a full occupancy. Taking into account the average length of stay in the particular subacute care program, you can compute the global number of patients required to keep the unit at a reasonable census. Additionally, it is ideal to have two beds available at all times to give referral sources access to the subacute care program.

As time passes, marketing staff should compare the variances in actual admissions to the goal. This type of analysis would have application to the revenue forecast.

New Service Development

This category of the marketing plan identifies the needs in the community. A new service campaign should create awareness and recognition. The future implications for the program should include the growth cycle and facility revenue when fully operational. This category is beneficial because it identifies those services that will make up a good percentage of revenue into the future.

IMPLEMENTATION AND MONITORING

Implementing the plan requires the entire staff's support. Aspects of the plan should be identified for participation by key department heads and appropriate staff. The amount of coordination with the various team members and department heads directly correlates to the ambitious nature of the plan.

Monitoring dictates that the plan must be considered a working document. It is referred to often and serves as a guide. It allows a subacute care program to monitor variances from goals. As in any endeavor, its predictive value becomes more limited as the forecast time frames are expanded. Unforeseen circumstances may require changes and adaptations to the components of the plan. Statistical trending can be useful in improving the predictive value of the plan; however, environments can change quickly and may not necessarily follow historical patterns. In general the plan should identify milestones and time frames for formal changes and additions.

The need for planning for a comprehensive marketing program is clear and requires attention to build consensus. Senior management's support and reinforcement of the goals increases the efforts put forth by the employees. This process serves to generate enthusiasm and galvanize employee energy for marketing among the staff.

PRODUCT LINE

Typically, as services develop, the need exists for the marketer to focus attention on awareness and understanding of this complex product. Recognition will greatly enhance the image of the provider. Helpful to this process are

- A dedicated name and logo for the subacute care program
- A position line
- A summary of the subacute care services for press releases and other media activity

- Consistent internal and external signage
- Various types of collateral materials
- Sales promotional tools

All of these items taken together will increase the brand equity that's built up in the subacute care product line. The most successful subacute care programs have built a high degree of brand equity over time.

As the competition intensifies, the provider will find itself in the maturity phase, where competition is high and a great deal of attention is directed toward price and volume concessions. Here the ability to differentiate and add value to the service will allow the provider to secure a return on investment that is above average. Winning market share becomes a determining factor in improving the competitive position of the provider. It is in this phase that the need exists to improve or develop new services and identify new customer segments. Advertising and public relations geared toward accentuating clinical competence will support the marketing efforts.

Tracking and identifying admissions by diagnosis is also important. For example, referral sources may refer mostly rehabilitation patients rather than medically complex patients. The use of subacute care for rehabilitation needs is better understood than needs of medically complex patients. Education and marketing should focus on building a greater understanding by referral sources of all types of subacute care. The subacute care rehabilitation patients can be used as a platform to attract other categories of patients.

A subacute care program can benefit greatly from the good will established by the parent corporation or facility. It can lend credibility and recognition to the subacute care program.

MARKETING PERSONNEL

The marketing staff should be viewed as the direct link with customers. Recently greater emphasis has been placed on the promotion of products and services of the healthcare industry. Many organizations have developed career ladders in order to accentuate the vital role of the marketing staff.

Persons with a clinical background are often prepared to handle the direct external marketing functions. A set of experiences with patient care is helpful when a marketer calls on other healthcare professionals. The ability to do an on-site assessment of the patient establishes the marketer as a knowledgeable resource.

Marketing skills normally are developed either through hiring experienced people or training. However, it is difficult for people from other industries to learn the healthcare system without some prior clinical experience.

Assuming that the strategy, market plan, budget, and image have been developed, the next step is to promote the subacute care product. Developing referral sources includes identifying key accounts, insurance case managers, referring physicians, social service departments, and discharge planners.

Remembering that 90% of the facility's business is referred from 20% of its referral sources, marketers should focus attention on the key accounts. Account management is critical to long term success. This involves identifying the attitude as well as current realities and perceptions of the customer, and the potential for referrals to the subacute care program. Structure activities to nurture the relationship over time through the initial meeting and tour. Introduce key members of the subacute care team and discuss the services. Provide information on the patient's progress and acquaint the account with new services. It is crucial to monitor referral results so marketing staff and other managers can evaluate strategies and make adjustments where necessary.

The marketer should understand that converting a referral into an admission requires consensus from many parties. The family, physician(s), social worker, case manager, and payor, in addition to the patient, may all be involved in making disposition decisions.

Therefore, networking becomes a crucial component of account management strategy. This process can be time consuming. In general, there is always more for the marketer to do than time allows. There may be calls in the evening from families, reports to complete for management, exhibiting opportunities to coordinate, or tours and special events to plan all in the course of a day. A marketing person should be able to handle multiple tasks. A skilled marketer will facilitate access to key referral sources. This access will provide valuable information related to market research, account intelligence, customer service, reputation, image, and community perception. The good will established will set the tone for the entire account relationship.

INTERNAL MARKETING

An organization cannot have an external system unless it has an internal system. Organizations may have limited knowledge and understanding of what it takes to compete successfully in the subacute care industry. With the arrival of managed care, the need to process referrals expeditiously and effectively cannot be underestimated. Most hospitals, and some subacute care programs, have moved toward an increased utilization of critical pathways. These pathways guide the course of treatment. They frequently prescribe a limited length of stay and require hospital discharge planners and social workers to be much more aggressive in moving higher acuity patients to subacute levels of care.

In consideration of the need to admit patients seven days a week, 24 hours a day, the facility must develop an internal marketing team. This team can be structured in many different ways and must acknowledge the need to treat referral sources as customers. Given that the majority of subacute care admissions come directly from the hospital, these relationships are crucial to any subacute care program.

The administrator has been accountable for all activities within the facility, traditionally. However, with increasing demands and a wide range of responsibilities, the administrator must delegate and develop some of the marketing responsibilities. One approach is to designate available personnel such as the case manager, director of nursing, or various other department heads to assist and facilitate the process of admission. The guiding principle that must be underscored should be to process referrals quickly and respond to the referral sources with a sense of urgency.

Important features of a marketing program include:

- Receptionist coverage should begin as early as possible in the morning and extend into evening hours. The receptionist is the first line of defense and sets the tone for the facility. An appropriate standardized greeting should be developed to ensure a warm, friendly initial contact.

- The clinical coordinator for the subacute care program or the director of nursing should be involved in reviewing the patient assessments and other data to discern the clinical needs of the subacute care patient. These individuals should have backup during times of vacation and absence so that the referral system continues to operate effectively.

- Enough telephone lines must be available so that the caller does not receive a busy signal, which can lead to frustration and loss of referrals.

- Adequate coverage should be appropriated so that evening and weekend referrals and admissions can be processed. This may require that the internal case manager or other staff members be on call and/or carry a beeper on weekends. Frequently, communication must be almost immediate between referral sources and the facility.

- Physician availability should be ensured for evening and weekend call as well. As managed care agencies seek to reduce expenses in the form of hospital days, it is incumbent on the subacute care program to evaluate and admit patients during the weekend. Active involvement by the medical director or attending physician of the subacute care program can improve the capability of the program to admit patients in relatively short time frames.

- The tour of the facility/program can be critical in determining success. Subacute care program tours will frequently take place during nonbusiness hours. Staff that have been inserviced and understand the importance of the tour become a key element of the marketing process.

♦ Customer satisfaction surveys are also beneficial and should be completed while the patients are in the building, as well as a survey tool to be evaluated upon discharge and follow-up. Additionally, a daily meeting should be held to coordinate activities as well as discuss admissions and discharges. This meeting is an important way to communicate across departments, create value, and present a coordinated program to customers. The ability to work together as a team enhances smooth and efficient functioning of the subacute care program. This might prove to be important during a tour, when people may be identifying and comparing one subacute care program with its competitors. Tours should seek to demonstrate the critical competencies of the facility, such as therapeutic recreation; physical, occupational, and speech therapy; expertise of nursing; comfort and convenience; physical plant and resources; and friendly and concerned staff.

A central 800 referral line can be efficient for one-stop shopping and making contact, particularly for insurance case managers. People who are operating the referral line must demonstrate the ability to communicate and educate effectively over the phone. This resource is valuable and gathers information that the facility requires in order to complete evaluations, and screen and admit patients. There should also be an evaluation tool that measures acuity, which will assist in pricing cases for managed care as well as identifying the facility resources required to adequately provide care and treatment to the prospective patient.

Knowledge of state regulations and requirements are essential and vary dramatically from state to state. Reviewing items such as state waiting list laws and work with patient advocates and ombudsmen can generate some understanding of the protocol within the subacute care program. There are issues about notification of patients regarding room changes, as well as providing criteria that explain when a patient is no longer appropriate for the subacute care program. Case managers, as well as facility social workers, can play an important role in developing an understanding. In particular, subacute care programs may be regarded as short stay, and this expectation should be developed or dispelled early on with the patient and family. This will improve coordination with managed care case managers.

PRICING

A chapter on marketing would not be complete without some discussion of the development of a pricing mechanism for cases funded by third-party payors. As a subacute care program evolves and matures, the opportunity to attract cases that are funded through the insurance industry increases. These cases are important to the success of the organization as it relates to financial performance and community image.

The goal is to cover costs and contribute to the profit margin. Identifying costs in terms both of direct costs including labor, as well as of indirect overhead expense, should be factored into pricing negotiations.

A number of subacute care competitors seek to differentiate themselves to some degree. These distinctions are difficult for the insurance case manager to identify. The case manager may believe that the available choice of providers offer a similar level of quality. The decision will then depend on location and access for the patient and family, as well as on price. Case managers from the insurance industry are charged with the mission to achieve tangible savings while providing the appropriate treatment setting.

In most cases, a fee-for-service discount is the least risky pricing method for the provider. It is simple to administer and ensures that the provider will be reimbursed the charge per unit of service less a percentage discount. The insurance case manager is less likely to be in a position to project the average daily cost to the carrier. Tangible cost savings accrues through the percentage discount negotiated. Most case managers are skillful negotiators and maintain close contact with the facility to monitor care and progress toward treatment goals.

In contrast, per diem rates may involve greater risk for the provider contingent on inclusions within the per diem rate and the provider's ability to understand and identify costs. Typically, the provider seeks to limit inclusions in the per diem rate, whereas the insurance carrier seeks an "all-inclusive" per diem rate. The predictive value is beneficial in terms of measuring savings compared to acute care. It also eliminates price variances in treatment fluctuations. A provider should negotiate per diem rates that are less comprehensive by the use of exclusions. These exclusions limit the provider's exposure.

There are still variations nationally and regionally in average per diem rates. Providers benefit from per diem rates to the extent that they have control of their cost structure and understand the negotiating factors within their catchment area.

Providers should renegotiate a per diem rate with the case manager upon status change of the patient's condition and treatment needs. Case managers expect the provider to reduce the per diem rate if levels of care are substantially reduced. If contractual rates have been prescribed, there is usually no need for extensive negotiation prior to admission. Normally, three or four different per diem rates are established based on patient clinical criteria and services offered. Often the only decision is which of the rate levels is appropriate to the patient's level of acuity.

More recently, capitated programs have been discussed with subacute care providers. These programs allow a provider to receive a fee based on subscriber enrollment in their catchment area. Specifically, the managed care organization will contract with a provider to handle a volume of subscribers at a fixed reimbursement per subscriber per month. It then becomes the provider's responsibility to care for the patient within the confines of the capitated rate. Eventually, capitation will likely be the payment method of choice by managed care organizations nationally.

The last payment type is the case rate. A subacute care provider is paid a lump sum to care for the patient based on expectations that are specific to the diagnosis. The subacute care program may choose to provide all or part of the services or to contract with other providers for some of the services. For instance, a provider may receive $30,000 to care for a patient who has suffered a C.V.A. To adequately provide services and make a profit depends on numerous variables and implies great risk. If the patient suffers complications, it may not only diminish any profits but may ultimately create a financial burden for the provider.

CONVERTING DATA TO INFORMATION

It is generally recognized that the most valuable companies in the world have built a high degree of brand equity in their product or service. Their product or service is highly valued and preferred over competing services. To some degree, customers will pay more for additional perceived value. This increased share of customers is derived by the skillful management and communication with the customer to create the desired image.

The subacute care industry has few established databases to profile and segment customers' needs. Few mechanisms have been developed to identify key indicators among potentially large accounts.

Data necessary to adequately manage a key hospital account, for example, might include the payor mix, DRGs, referral-to-admission ratio, referral source, physician utilization, unit of the hospital most referred from, amount of subacute care referrals made by a hospital per month, percentage received by the subacute care program, and the reputation of the subacute care program as perceived by the hospital.

These types of data should be contained for each key account profile. The information should be part of a database system and accessible for input and retrievable by various members of the subacute care team. The relationship should be developed and tracked over time. This information can be used to develop account strategy and identify future potential. It becomes a useful tool to assist in managing the relationship.

A weekly report that records key indicators and allows for trending and analysis of significant marketing activity should be completed. Historical profiles can be developed with key indicators such as

- ◆ Referral-to-admission ratios
- ◆ Marketing call-to-admission ratios
- ◆ Marketing activity
- ◆ Payor mix summaries
- ◆ Discharge activity
- ◆ Admissions by referral source

Other data include mailing lists coded by target audience, insurance contract profiles and renewal dates, and key account activity. This information must be captured in the appropriate format to monitor results versus goals. It can be used as a management tool and indicator of the effectiveness of the marketing effort.

ETHICAL MARKETING PRACTICES

Any discussion of marketing should include a dialogue on ethics and ethical marketing practices. In this age of consumerism, ethics are a particularly important subject. Representations and assertions can later be challenged and can influence prospective patients and families. Training and communication should be implemented concerning ethical behavior from the organization's perspective.

Ethics in a business context as it relates to healthcare presents organizations with issues of social responsibility and development of company values. Values and principles guide decision making and to a large extent dictate actions of employees. These rules that govern employee behavior will be evaluated by referral sources and other key influences in day-to-day activities of marketing. In most situations people have differing perceptions of particular issues. Ethical values that are developed and practiced by programs/facilities and companies set a standard and improve decision making within the organization. In other words, it is important to do the right thing at the right time for the patient.

Pressures to generate revenue and fill beds often require the marketer to "move with a sense of urgency." Most people performing marketing tasks are articulate and enthusiastic by nature. It can be natural for these people to create hope and generate high levels of credibility with referral sources. This approach must be tempered with realism and an understanding that people are vulnerable when faced with traumatic events and difficult decisions. Most often, families are thrown into situations that they are ill prepared to handle. In their efforts to do something for their family member or loved one, families will try to gain knowledge to better understand the treatment regime and learn about the expertise of the subacute care program staff. The marketer or clinician should be concerned about the accuracy of any representations.

Attention to presenting honest, accurate, and achievable outcomes is important. Staff should not promise or create unrealistic expectations on the part of the customer. This will damage credibility with the patient and family. The facility may be unable to meet these expectations during the patient's length of stay.

Often, patients and families will have other programs and/or facilities to choose from. In fact, in some urban areas there are normally two or three organizations that have performed on-site visits to evaluate the patient prior to a family making a decision about choosing a facility. Competitors should not discuss the competition. Each facility should represent their services accurately and openly and allow the customer to compare those services.

A facility may have unique clinical competencies that differentiate it from other centers. Competencies can be reflected in brochures and other collateral materials such as outcome data. This is helpful and can be provided accurately in marketing literature. These materials should be evaluated frequently and always when services change. Including a third party in this evaluation will allow for an objective opinion and assist with understanding how a customer is likely to interpret the information. It is also important not to create unrealistic demand by customers for services. This is particularly important in healthcare when third-party payment is involved. Marketers should always identify that their comments are general in nature and cannot be specific until the individual patient's circumstances are known. In particular situations marketers must make certain that in performing patient evaluations, they are qualified and trained to assess the clinical aspects of the patient's needs. Normally, hospitals have procedures for completing on-site assessments.

SUMMARY

The organization or healthcare provider should seek to develop a set of guiding principles for the subacute care program's marketing force as it relates to all marketing issues. The ability to develop these values early will assist consumers and referral sources, and enable the marketer to identify program or company values. This value identification is important to success and should form an integral part of an organization's marketing program.

The marketing effort and the staff members that implement it have become two of the most important components of any organization. A plan, ample appropriation of funds, and qualified personnel are vital to a quality subacute care program.

CHAPTER 10

Outcomes

'It was with hope and triumph that I ventured in my new shape."

— Dr. Jekyll

OUTCOMES MEASUREMENT

In general, *outcomes measurement* applies to the measurement of the change in the health status or functional status of the patient attributable to healthcare provided. In the context of managed care, it also applies to the concept of providing the best quality services at the lowest possible cost and determining the point at which quality will not be sacrificed if fewer resources are used. Paul Ellwood at the 99th annual Shattuck Lecture to the Massachusetts Medical Society described outcomes management as a method of helping patients, payors, and providers to make reasonable medical care choices based on better insight into the effect of these choices on the patient's life. In addition, the Clinton Administration proposed in the Health Security Act in 1993 to fund research that would assist in answering questions about what treatment works best for which conditions, so that physicians can provide the highest quality care for their patients.

The burgeoning subacute care industry is currently providing service for both rehabilitation and medical patients, and often the same patients are receiving both medical and rehabilitation services. Therefore, subacute providers must consider a broad spectrum of outcome measures to cover diverse programs.

Outcomes can be described in terms of the patient, a program, or an episode of illness. Patient outcomes apply to the outcome of medical treatment or rehabilitation on a single patient. This usually applies to the concept of the patient meeting his or her treatment goals. Program outcomes report the effectiveness and efficiency of a specific program, such as a subacute care program. There is increasing interest in measuring outcomes of a variety of interventions applied across an episode of illness. This episode may involve a variety of programs, from acute to postacute care. In this case the emphasis is on the mix and amount of services provided to a group of patients. Further research is needed on the outcomes of an episode of illness. The focus of this chapter is on a program, subacute care; therefore, this chapter will focus on program outcomes rather than patient outcomes or the outcomes of an episode of illness.

EFFECTIVENESS OUTCOMES

Effectiveness outcomes should be measured from the perspective of the client as well as the healthcare professional to achieve a full picture of the outcome. Measures of effectiveness include

- ◆ Medical improvement of clients during a stay in a program or from onset to resolution of an episode of illness
- ◆ Functional improvement of the patient during a stay in a program or from onset to resolution of an episode of illness
- ◆ Patient's perception of his or her health status and quality of life
- ◆ Patient satisfaction

Medical Improvement

In subacute care, the industry wants to determine whether similar medical outcomes can be achieved in various settings, such as a subacute care setting, an acute care setting, or a home healthcare setting. Also of concern is whether one subacute care program provides comparable care to that provided by other subacute care programs in a given region. To make such comparisons, the status of patients having the same diagnoses and impairments but treated in different settings must be measured at the patients' admission and discharge from each setting. The settings must agree to employ identical methodologies to record the same data for comparison purposes. Some subacute data collection instruments are in the research phase, so it will necessitate compliance and cooperation in such an effort. The subacute care industry is driven to prove its efficacy and appears willing to put forth the effort to collect the necessary data. Whether other areas of healthcare will permit themselves to be compared on a voluntary basis remains to be seen.

The major medical areas of focus for subacute care services, and thus outcomes measurement, are:

- ◆ Wound care
- ◆ Pain
- ◆ Antibiotic and nutrition therapy
- ◆ Respiratory care
- ◆ Infusion
- ◆ Postsurgical services
- ◆ Monitoring
- ◆ Cardiac care
- ◆ Renal management

In all cases, the positive outcome would be medical improvement. For example, in wound care, the outcome could be that the wound was healed, the wound decreased in size, and/or the infection was resolved. An additional outcome might be that the person with the wound could live at home, rather than in a facility and no longer required medical care.

Functional Improvement

Basic functional status applies to a person's ability to perform self-care activities known as *activities of daily living (ADL),* his or her ability to ambulate or use a wheeled mobility system, to communicate, and to use cognitive skills. Also included in this category are what is commonly referred to as *instrumental activities of daily living (IADLs)* that include homemaking, financial management, travel, driving, and performing leisure activities. One might also consider the ability to work as a higher level functional activity. Function is measured by the clinician's direct observation of the patient attempting to perform the activities in question.

One challenge in the measurement of function is to develop a conceptual framework. In 1960, Nagi developed the following framework for relating disease, impairment, functional limitations and disability:

- ◆ *Pathology* represents interruption of, or interference with, normal bodily processes or structures. Pathology involves abnormalities at the level of tissues or cells, such as denervation of a muscle in the arm by trauma.

- ◆ *Impairment* reflects abnormalities or losses of function along specific anatomic, physiological, mental, or emotional dimensions. By definition, all active pathology involves impairment, even if transitory. However, impairments can also result from causes other than active processes, such as congenital malformations, and residual impairments can remain following resolution of an active pathologic process. Impairment considers the level of organs or organ systems, such as atrophy of the arm.

◆ *Functional limitations* represent restrictions or inability to perform various activities considered normal because of impairments. Limitations focus at the level of the person. An example of this is the inability to use the arm for tasks involving pulling.

◆ *Disability* involves limitations or inability to perform activities and roles defined by society and the external physical environment. Disability therefore focuses on various aspects of the social and cultural context. For example, a person suffering an arm trauma may no longer be able to work, thus representing a disability.

Research has indicated that the best predictor of a patient's future functional status is his or her current functional status. Therefore, the measurement of a person's status at admission is the key indicator of the discharge outcome.

In measuring functional outcomes, the investigator must adhere to a conceptual framework to develop a way in which to classify patients measured by the instrument. The customary method of classification of rehabilitation patients is by impairment category. Once a category is selected, the level of disability is measured using scales that rate the person's level of disability.

Patient's Perception of Health Status and Quality of Life

Momentum has been building for the acknowledgment and inclusion of patients' perspective of whether there was improvement in the quality of their lives as a result of receiving healthcare services. This assessment includes the patients' mental health, physical functioning, and role functioning in their social and work setting. Some administrators would go so far as to say that patients' perspectives are really the only outcome that matters. However, it is important to recognize that patient attitudes and perspectives change through time and circumstances. Therefore, outcomes researchers would prefer to include both the clinical and patients' perspective.

Patient Satisfaction

Along with an emphasis on patients' perceptions is the continued importance of patient satisfaction. Patient satisfaction is considered to be the difference between the patient's expectations at the time of implementation of therapy and the patient's perception of the services received and therapeutic results. Patient satisfaction criteria should include an assessment of both clinical and nonclinical parameters. Clinical parameters might include a determination of whether the patient thought the intervention improved his or her abilities to function, or decreased pain. Nonclinical but clinician-related factors include an appraisal of the client's perception of the empathy of the staff and the staff's communication skills. Nonclinical parameters might focus on structural components of the service delivery, such as comfort of the facilities, timeliness of therapy sessions, and convenience of appointments.

EFFICIENCY OUTCOMES

Efficiency outcomes are intended to measure resource use and the economic factors associated with outcomes. Measures of efficiency may include

- The length of stay (LOS) for inpatients.
- The length of service for outpatients: time from admission to discharge.
- The number of treatment units or visits by healthcare professionals for inpatients and outpatients.
- The average charges per visit or the net reimbursement after discharge adjusted by the payor.
- The number of full-time equivalent staff (FTEs) used in the patient's care. This can be further delineated to examine the mix of licensed professionals and the number of assistants or aides used in the patient's care. In rehabilitation, one may also wish to track the mix of individual and group treatment sessions.
- Utilization of resources after discharge. In other words, what healthcare services continue to be needed after the patient is discharged from the program.

Outcomes are the interaction of structure and process standards. *Structure standards* include such factors as the composition and number of staff, the type of equipment, and the layout of facilities. *Process standards* refer to procedures such as treatment techniques, the interaction of the team, and the ability of therapists to involve clients in their care. Process standards are commonly referred to as *critical pathways* or *clinical pathways* and must be considered when seeking to improve outcomes. When structures and processes are combined, outcomes are produced. The measurement of outcomes does not directly involve the measurement of processes or treatment techniques; that process is known as *clinical assessment.* Many programs make their outcome measurement system more complicated than it needs to be by attempting to measure far too many clinical assessment measures. It is important to remember that outcomes are the results of the program as a whole, not the individual client or steps involved in achieving the outcome. It serves as the report card of the overall results of the program—the big picture.

SOURCES OF OUTCOME DATA

The ideal source of data would be one that was completely accurate, contained reliably measured medical or functional gain, and required no extra time from clinicians to gather. Sources of actual data regarding outcomes are

- *Administrative data*—large, computerized data files generally compiled in billing for healthcare services, such as hospitalizations usually obtained through government and insurance industry sources
- *Medical record information*—data elements abstracted usually retrospectively from the medical record or some other source (such as computerized laboratory reporting systems) and obtained from the healthcare provider
- *Patient-derived data*—information collected directly from patients either through interviews or questionnaires
- *Program evaluation data*—In the rehabilitation industry, outcomes data are commonly available through program evaluation instruments
- *Minimum data set (MDS) data*—an extensive data collection instrument that Medicare plans for facilities to use; MDS requires data be collected on both medical and rehabilitative parameters as they apply to patients receiving long-term care

Administrative Data

With the increased emphasis on patient outcomes, most studies and reports have been based on administrative data sources because they were available and relatively easy to access. The hospital mortality data for Medicare beneficiaries, published by the Health Care Financing Administration (HCFA), is the most well-known example of an outcome study based on administrative data.

The Uniform Hospital Discharge Data Set (UHDDS) was the first administrative data set developed. This data set includes 14 core data elements from all short-term hospital discharges paid through Medicare and Medicaid. These elements include basic patient demographics, physicians identified by unique numbers, lengths of stay, procedures and dates, patient disposition, and expected pay source. The UB-92 (Uniform Bill first introduced in 1982) has 85 data entry fields. The UB-92 data are fed into a database that includes Medicare, Medicaid, Blue Cross, and many commercial carriers. The HCFA Office of Statistics and Data Management has data files that may apply to the skilled care and subacute care industry.

Additional administrative data files of variable quality are available through state Medicaid files and private insurance claims.

Medical Records Data

As states have become more interested in monitoring healthcare outcomes, some have demanded more information on severity than is available through administrative data sources. This effort was motivated by a concern about the accuracy of administrative data sources and the fact that patient severity was not considered when reporting information such as hospital mortality rates. One method of compensating for this problem is through abstracting clinical information from the medical record. The problem is that the accuracy and completeness of medical documentation varies between facilities. Functional and social information, pertinent to outcomes in subacute care, is commonly missing.

Patient-Derived Data

The impetus for the need to consider patient perception in healthcare outcomes has resulted in a deluge of newly developed health status and quality of life measures for most impairments. A popular health status measure is the Health Status Questionnaire 2.0 created by the Health Outcomes Institute. This is a written questionnaire that consists of 39 questions, surrounding the topics of mental well being, physical functioning, social functioning, and role functioning. The questionnaire may be completed by the patient or by a trained interviewer.

In subacute care, patient-derived data can be difficult to obtain and unreliable, particularly if the patients have cognitive or sensory deficits making it difficult to answer interview questions or to complete a questionnaire. A response for this dilemma would be to shorten questionnaires to ask only the most pertinent questions.

Program Evaluation Data

In program evaluation, clinicians rate patients' level of function at admission and discharge in order to evaluate the need for programs and their effectiveness. For most commercially available systems, the clinicians are certified to rate patients on functional assessment scales for the purpose of achieving interrater reliability. Program evaluation data are collected concurrently and although subject to rater subjectivity, are probably more accurate and complete than the sources described previously. The quality of these data must be balanced with the time and expense of data collection.

Two functional assessment scales in common use that are connected to national databases and are marketed commercially include the Functional Independence Measure (FIM) of the State University of New York, National Institute on Disability and Rehabilitation Research (1984); and the Level of

Rehabilitation Scale or LORS-III by Carey and Pasavac (1978). Detailed information on each of these scales is provided later in this chapter.

The MDS

The MDS is completed on all long-term care patients at admission and every 90 days. Patients are classified for the purpose of Medicare payment. Another objective for developing the MDS was also to monitor the quality of the process and outcomes of care and adjusted for case-mix. It is also intended to identify patient needs where special protocols should be implemented. The MDS scores patients in the following categories:

- ◆ Section A. Identification and background information
- ◆ Section B. Cognitive patterns
- ◆ Section C. Communication/hearing patterns
- ◆ Section D. Vision
- ◆ Section E. Mood and behavior patterns
- ◆ Section F. Psychosocial well-being
- ◆ Section G. Activity pursuit patterns
- ◆ Section H. Physical functioning and structural problems
- ◆ Section I. Continence in last 14 days
- ◆ Section J. Skin condition and foot care
- ◆ Section K. Disease diagnoses/health conditions
- ◆ Section L. Oral/nutritional status
- ◆ Section M. Oral/dental status
- ◆ Section N. Special treatments, devices, and procedures
- ◆ Section O. Medication use
- ◆ Section P. Participation in assessment

Researchers have studied the potential efficiency of employing MDS data for outcomes and have discerned that the instrument poses several problems. First, the MDS would need to be completed on patients at admission and at discharge and a method of analysis would have to be developed to determine all patients' medical and functional progress. At the initial review, it seems that the MDS might provide the basis for an outcomes measurement system. Sections from the MDS could be scored at admission and discharge, and additional outcome variables and more detailed definitions could be added. However, after further study, it becomes apparent that many of the items that seem to provide a basis for outcomes measurement are not designed to measure significant patient progress. For example, Section H, which relates to physical functioning and structural problems, includes items such as bed mobility, transfer, locomotion, dressing, eating, personal hygiene, and bathing that apparently serve as a foundation for a functional assessment scale for patients receiving rehabilitation. A challenge to using MDS for outcomes is that patients are scored on two different constructs: both on ADL self-performance (independence) and on ADL support (burden of care). Researchers have considered either combining and weighting the ratings or using only the ADLS self-performance measures for outcomes. Neither strategy adequately responds to the question of "What outcome did the patient achieve?"

Section J, which relates to skin condition and foot care, seems to provide a basis for measuring outcomes in wound care, but clinicians view the scale as insensitive to change. This same issue seems to apply to every section of the instrument that would seem to apply to outcomes. Further research is necessary to improve the utility of the MDS for outcomes measurement.

SELECTING OR DESIGNING AN OUTCOMES MEASUREMENT SYSTEM

First you should identify the users or customers of your outcome data and find out what they want to know. The major users of your information will include administrators, clinicians, payors, employers, and consumers. In general, administrators are concerned with improving quality and using the information for marketing to payors. Clinicians tend to be interested in specific, clinical assessment measures for specific diagnoses that show small increments of change. Payors are interested in the overall program performance for categories of patients. Payors and employers are interested in comparing the performance of facilities. As a result, it is desirable for all facilities to use the same method of measuring outcomes. Payors are also interested in predicting resource utilization of patient categories to determine overall cost projections for managed healthcare. The outcome data can then form the basis for a predictive model. Patients assess quality through a perception of caring more than by outcomes. They are interested in knowing whether the therapist explained in advance what the therapy would involve and what results could be expected. Patients want to know that the physician expressed concern about their particular situation. To most patients, a positive outcome occurs when the rehabilitation experience matches or exceeds their expectations. Outcome data can be used to educate patients about what to expect when they begin receiving rehabilitation services.

Although a group of expert clinicians can determine logical steps for patient improvement in most areas, it is difficult to gain consensus among a number of facilities that the outcomes that are being measured are those of concern and that the steps defined by the expert panel are accurate. Scales must be analyzed for validity and reliability. It is important that the scales fit the ability of the persons being measured. Also, if outcomes cannot be compared among facilities, they lose their meaning to payors and consumers, becoming simply an internal quality improvement tool.

MEASUREMENT OF REHABILITATION OUTCOMES IN SUBACUTE CARE

It has been estimated that a large percentage of the subacute market consists of providing services to physical rehabilitation patients, such as those who have experienced strokes and hip fractures, and the rest of the subacute market consists of medically complex patients, such as patients using ventilators or receiving infusion therapy. Many patients are difficult to classify into one of these categories exclusively as they may be receiving rehabilitation services and may have a medical diagnosis.

Those providing subacute care have an interest in comparing their outcomes to those of traditional services. The comparison is necessary to demonstrate efficacy of treatment in this less expensive setting. For subacute rehabilitation, the industry has sought a method to compare outcomes to acute inpatient rehabilitation that occurs in freestanding rehabilitation hospitals or in DRG-exempt units within acute hospitals. The subacute industry has borrowed program evaluation instruments that have been commonly used in these acute settings. The Functional Independence Measure (FIM), used by over half of the acute care settings in the United States, is commonly used in subacute settings. This scale has been validated in acute rehabilitation. More study is required to be certain of its validity in the subacute setting.

An observation that lends credence to the transference of validity to the subacute setting is that many of the same patients who have been seen in the acute setting are now receiving treatment in the subacute setting. Preliminary research is underway to determine the effect of medical comorbidities and their severity level on functional outcomes.

Functional Independence Measure (FIM)
The Functional Independence Measure was developed in 1984 by a joint task force sponsored by the American Congress of Rehabilitation Medicine (ACRM) and the American Academy of Physical Medicine and Rehabilitation under a three-year grant from the National Institute on Disability and Rehabilitation Research. The grantee institution was the State University of New York at Buffalo located at Buffalo General Hospital. Carl V. Granger, M.D., was the project director and Byron B. Hamilton, M.D., Ph.D., was the principal investigator.

The objective of the task force was to develop a patient data set and an assessment instrument to be used universally among all inpatient rehabilitation units.

The FIM rating is based on a seven-point scale, shown in Exhibit 10.1, that reflects a philosophy of burden of care or the ability to quantify the amount of care or assistance required. The rater must first decide whether the patient is

---

EXHIBIT 10.1
FUNCTIONAL INDEPENDENCE MEASURE

Functional Independence Measure (FIM)

Self-Care

Eating
Grooming
Bathing
Dressing—Upper Body
Dressing—Lower Body
Toileting

Sphincter Control

Bladder Management
Bowel Management

Transfers

Bed, Chair, Wheelchair
Toilet
Tub, Shower

Locomotion

Walk/Wheelchair
Stairs

Communication

Comprehension: Auditory and Visual
Expression: Vocal and Nonvocal

Social Cognition

Social Interaction
Problem Solving
Memory

---

independent in performing the task and whether the patient requires the utilization of equipment to be independent. If the patient is not independent, the rater must decide whether the patient requires the assistance of another person to perform the activity and if so, whether the assistance provided is hands-on or supervisory only. Once it is determined that the patient requires hands-on assistance, the rater must decide whether the patient requires minimal, moderate, or maximum assistance.

FIM consists of a number of items and is intended to be discipline free. Clinicians who rate patients using the instrument are expected to be able to rate on all items. Often ratings are negotiated in a team conference. When there is a discrepancy, the lowest score or the most dependent observation is recorded.

The Level of Rehabilitation Scale (LORS-III)

The Level of Rehabilitation Scale (LORS) was developed in 1977 by Raymond Carey, Ph.D., Emil J. Pasavac, Ph.D., and Aaron Rosenthal, M.D., to meet a need in the rehabilitation field to evaluate and compare the effectiveness and efficiency of inpatient rehabilitation programs. It was based upon Sarno's Functional Life Scale. The objectives were threefold: (1) to provide rehabilitation staff with feedback on the amount of change they were achieving in the rehabilitation program; (2) to provide a method of satisfying legal and accreditation requirements with normative data; and (3) to determine which patients could benefit from physical rehabilitation. The instrument is interdisciplinary; however, each discipline is expected to score certain items at the time of admission and discharge. Two raters from different disciplines are required for every item and ratings are made independently, without collaboration with other disciplines. LORS is designed to be quickly administered. LORS was revised in 1980 and became known as the LORS American Data System or LADS. The database of more than 50,000 patient records was managed by Parkside Associates, Rehabilitation Research Division, a division of Lutheran General Health Care System in Park Ridge, Illinois. In 1991, ownership of LADS was transferred to Formations in Health Care, Inc., in Chicago. In 1992–1993 another revision was completed through adding more detailed cognitive scales, and the instrument became known as the LORS-III.

The LORS-III rating is based on a multilevel scale, converted to percentage scores as follows: 0 (for example, totally dependent), 1 (for example, 25 percent independent), 2 (for example, 50 percent independent), 3 (for example, 75 percent independent), and 4 (for example, 100 percent independent). The scale consists of 18 items, rated by dual raters on all but one item. A sample LORS-III rating appears in Exhibit 10.2.

EXHIBIT 10.2
SAMPLE LORS-III RATING

LORS-III and Designated Raters

	Nurse	OT	PT	SLP
Activities of Daily Living				
Dressing	✓	✓		
Grooming	✓	✓		
Washing/Bathing	✓	✓		
Toileting	✓	✓		
Feeding	✓	✓		
Mobility				
Ambulation/Wheelchair	✓		✓	
Communication				
Auditory Comprehension	✓			✓
Oral Expression	✓			✓
Reading Comprehension				✓
Written Expression				✓
Cognitive Ability				
Attention		✓		✓
Orientation		✓		✓
Problem Solving		✓		✓
Sequencing		✓		
Memory				
Short-term	✓	✓	or	✓
Long-term	✓	✓	or	✓

OT = Occupational Therapist
PT = Physical Therapist
SLP = Speech/Language Pathologist

There are some problems with using the LORS-III in subacute care. Although the LORS-III is less time consuming than the original LORS, it can be difficult to apply to a subacute setting. There is not always the opportunity to obtain dual ratings, or the patient may not be receiving care from all the disciplines required to complete the ratings on the scale.

Some therapists have suggested that transfers, swallowing, and bowel and bladder management be included although these skills are included in the definitions of other items on the scale. Several problems exist with both of these scales in their use with subacute patients. Most clinicians tend to find the areas of measurement not comprehensive of the treatment they provide and tend to find the scales less than sensitive in indicating patient progress.

Subacute care settings serve a diverse group of patients, and preliminary data indicate that patients receiving subacute rehabilitation services have more medical comorbidities and chronic conditions than the patients receiving rehabilitation services in hospital settings. It is not enough to simply track the existence of a comorbidity; it may be necessary to also classify the acuity. Further study is needed to determine the effect of comorbidities on this patient population.

Both scales have been tested for reliability with acute hospital rehabilitation patients but have not been generalized across alternate settings. Research has indicated that the same functional status measure may not perform equally well across the entire spectrum of a single disability.

Some shared features of both scales:

- Clinicians rate patients at admission and discharge according to their observation of the patient's ability to perform the activities in the scale. Ninety days following discharge, a telephone interview is conducted to compare the patient's functional level to ratings attained at discharge.

- Additional descriptive and demographic fields as well as length of stay, gross and net charges, and discharge disposition fields are added to both scales.

- Analysis of the fields determines program outcomes such as success in patients returning to work, success in discharging patients to the community or a lesser level of care, success in maintaining expected lengths of stay, and success in charging comparable rates as other rehabilitation programs for the treatment of patients with similar impairments.

MEASUREMENT OF MEDICAL OUTCOMES IN SUBACUTE CARE

The idea of measuring outcomes in subacute medical care is new to the industry, and no instrument exists that is commonly used by providers, although many providers have developed or are in the process of developing one. Research to develop such an instrument is underway by outcomes measurement companies. Instruments under development are intended to be used throughout the subacute industry and acute care and home healthcare. Some instruments attempt to predict improvement in medical and functional status rather than mortality or complications. One such instrument is being developed by Formations Health Care, Inc., in Illinois. A sample of items for medical outcomes appears in Exhibit 10.3.

ANALYSIS OF OUTCOMES

Rehabilitation Outcomes Analysis

Functional scales are ordinal scales; therefore, accurate reporting and analysis of these scales can be challenging. In the rehabilitation industry, two different methods of outcomes analysis are provided by the two companies providing reports based on national databases. For reports provided by the Uniform Data System, Buffalo, New York, outcomes information is reported by adding the score of all items to derive a composite score.

Formations converts the ordinal scale to an interval scale, so the data can undergo a complex data analysis. In this method, levels of function in each area are reported according to percentage of independence. Percentage gain for patients in each impairment group in a facility is compared to the average of the database and also to an expected range. The expected range is derived from an analysis of the database to determine a range of outcome adjusted by patient acuity, functional level at admission, region of the country, and facility type.

Medical Outcomes Analysis

There are several models used for analysis of medical outcomes and it is unknown at the time of this publication what methods will be accepted for Medical Outcomes Scales. In the APACHE II, an Acute Physiology Score (APS) is given. This score is generally accepted as an estimate of severity of illness in individual patients. In the Simplified Acute Physiology Score (SAPS), the scale accounts for the number of body system failures. The analysis used in these instruments, as well as the application of Rasch analysis, will be considered for the analysis of the Medical Outcomes Scale.

EXHIBIT 10.3
MEDICAL OUTCOMES SCALE ITEMS
(FIELD TEST II VERSION OF FORMATION'S OUTCOME SCALE)

Wound Program

Total number of wounds
Total volume
Perimeter involvement
Drainage
Depth of wound

Respiratory Program

Type of program
Resource utilization
    Oxygen
    Ventilator
    Tracheostomy
Health status measures
    Dyspnea
    Cough
    Consistency of secretions
    Breath sounds
    Oxygen saturation
Date of resolution for ventilator weaning/tracheostomy
    decannulation

Infection Program

Physiological responses
Severity of infection
System dysfunction associated with infection process

Pain Program

Pain scale
Duration of pain
Pain relief interventions
Type of medication

Functional Status

The functional independence measure or the LORS-III
items comprise this section at the discretion of the facility
using the instrument.

"RISK ADJUSTMENT" OF OUTCOMES

Administrators should avoid making comparisons of ratings to the average of a database. The accepted way to compare data of a single facility to a database is to use a method that Lezzoni called *risk adjustment.* Risk adjustment accounts for pertinent patient characteristics before assessors make a conclusion regarding the effectiveness of a facility. Risk adjustment is necessary because the patients who are admitted to a facility do not represent a random sample and may not necessarily reflect the condition of average patients.

One dimension of risk adjustment is referred to as *severity of illness.* Severity has been an important consideration in healthcare since the implementation of Medicare's prospective hospital payment systems based upon diagnosis related groups (DRGs). Generally, severity is of concern to facilities that tend to admit more severe cases compared to other facilities. Severity is associated with cost. Therefore, more reimbursement is expected for treating more severely ill patients. In rehabilitation severity, there is less concern with medical acuity and more concern with functional status as the outcome of interest. Facilities are concerned about accounting for patients who have lower levels of functioning at admission, rather than measuring severity or acuity.

One statistical method used in accounting for severity of illness in medical patients or lower functioning in rehabilitation patients is multiple regression analysis. By applying this analysis to an individual rehabilitation facility's data, outcomes (or the expected level of independence at discharge) can be predicted. This analysis takes into consideration what the facility's outcome should have been, given the variables used to rate the patients. In multiple regression analysis, outcome variables such as length of stay, charges, level of independence at discharge in self-care, mobility, communication and cognition, and return to home rate are predicted by including in regression equations those variables thought to have power in predicting outcome variables. The variables that have been found to have influence in predicting outcome include

- Percentage of independence at admission for self-care
- Percentage of independence at admission for mobility
- Percentage of gain for self-care
- Age
- Gender
- Living arrangement prior to rehabilitation admission
- Health Care Financing Association (HCFA) region
- Type of facility (hospital-based, freestanding rehabilitation hospital, subacute care program, skilled care facility)

- ◆ First admission (yes or no)
- ◆ Occurrence of major surgical procedure during last 60 days

MANAGING RISK FOR COST AND OUTCOMES

In a managed care scenario, a facility may be asked to guarantee a certain outcome for a predetermined price. To accomplish this, a severity of illness adjustment or classification must be used to be certain that outcomes are indeed achievable, and can be reached within a specific amount of time and with a specific amount of resources. Using the severity of illness adjustment, a range of outcome can be guaranteed within a 95% confidence interval. This means that 95 times out of 100, the patient's outcome should be within the guaranteed range.

OUTCOMES REPORTS FOR REHABILITATION SERVICES

Commercially available outcomes systems provide subscribers with reports that compare an individual facility to national data. These results are usually reported quarterly and include the outcome domains of length of stay, overall charges, and functional outcomes for activities of daily living, mobility, cognitive ability, and communication.

The results are intended to provide powerful information to clinicians, clinical managers, marketing staff, and administration. If a facility receives unfavorable results, the facility should study and document the processes used and develop a quality improvement strategy. If a facility receives good results, its managers should study and document processes and integrate these processes into critical pathways. Often facilities with good results will strive to exceed previous performance.

An organization's core marketing strategy can play a part in determining which outcomes will be reported. Outcomes can help the organization to position itself as a market leader, market challenger, or market follower. If the organization positions itself as a market leader, data can be publicized that indicate the facility's results are better in all categories and that it is providing additional services not available elsewhere in the region. An organization that is a market challenger will make every attempt to compare itself directly to the competition and will hope to demonstrate better outcomes. If the organization is a market follower, data should indicate that the facility is performing at least as well as the market leaders and that the facility is providing comparable services to its competition. An organization may collect more comprehensive and detailed data on a narrow segment of a single niche or a comprehensive program.

For example, an orthopedic program specializing in hip replacements might have more specific data that include the outcomes resulting from the application of certain interventions such as vitamins, surgery, therapy, and medications than those that might be measured in a typical orthopedic program. Facilities are often eager to publicize good results, but once they begin to publish results, consumers will expect continued publication, even if the data are not so positive. Positive outcomes can support a position to demand a more advantageous payment for contracts and can prove a program is more efficient and effective than others.

SUMMARY

Outcomes measurement in subacute care programs is in its formative stages. There are few existing administrative or medical records databases from which the industry can draw outcomes measures. The MDS seems better suited to patient classification for payment than for the measurement of outcomes. Consequently, subacute rehabilitation programs are borrowing outcomes instruments from acute care. These instruments need further development to consider the more medically complex rehabilitation patient. Research is underway to develop an outcome measurement system for medical care with the intent that it can assess outcomes across a continuum of acute, subacute, and home care programs. The industry can expect outcomes research to flourish along with the increased volume and variety of patients successfully treated in the less expensive setting of the subacute care program.

CHAPTER 11

Accreditation

"Our patients expect us to be accountable."

— Dr. Ben Casey

Standards for accreditation for subacute care facilities have been developed by the Joint Commission on the Accreditation of Healthcare Organizations (JCAHO), Longterm Care Accreditation Services Division (LCASD), and by the Commission for Accreditation of Rehabilitation Facilities (CARF).

The process of developing these standards began a few years ago and was prompted by several factors. It is important to be able to distinguish the difference between the varied levels of healthcare. Defining subacute care will promote a clearer understanding of its utilization and practice criteria. Managed care companies, recognizing the cost effectiveness of subacute care, requested standardization of provider services. Case managers sought a method for quantifying outcomes resulting from placement in a specific level of care. Consumers were becoming more sophisticated in both their knowledge of healthcare and its costs. This led to requests for information related to healthcare purchasing choices. Additionally, the federal government and the individual states' reimbursement systems were incurring increasing amounts of debt due directly to the burgeoning expense of providing healthcare. Facilities were striving to gain recognition and reimbursement for subacute care programs.

JCAHO and CARF began the process of performing extensive research. This included surveying the market, reviewing standards already established for acute hospital care, long-term care, and acute rehabilitation. The results of this research indicated a need for subacute care standards of accreditation. Expert panels were established by CARF and JCAHO. The two panels consisted of professionals representing physicians; nursing administrators; case managers;

physical, occupational, speech, recreation, and respiratory therapists; providers from multisystem organizations as well as individual facilities and nonprofit and for-profit corporations; pharmacists; social workers and psychologists; dietitians and nutrition therapists; and payors, including managed care companies. There was also representation from patients and family members. Both of these panels worked to complete the task of creating the criteria and standards that JCAHO and CARF would use to survey subacute care programs.

This chapter contains information that will assist those providing subacute care in becoming accredited by removing the mystery surrounding many of the expectations. It will give the payor a sample of minimum standards and will promote a better understanding to enable the consumer to make educated choices.

CARF: THE PROCESS

The initial step in the accreditation process is for the organization to contact the Commission for Accreditation of Rehabilitation Facilities. (1) The Medical Rehabilitation Division will send information on the process of accreditation and how to purchase a standards manual, assign the facility contact people at CARF, and give information about any meetings or conferences that are being held to assist the facility in the preparation for a survey site visit. (2) An organization that has not sought accreditation previously needs to spend approximately 12 to 16 months in preparation from that initial phone call to the completion of the survey process. (3) The survey is performed by peers who are from CARF-accredited facilities and generally lasts for a period of two days. (4) CARF sends two surveyors, an administrative surveyor and a program surveyor. (5) The survey is a combination of observations, interviews, and interactions with consumers, staff, payors, and various other interested parties. (6) A review of records, policies, and procedures is conducted. (7) The results of the survey are received four to six weeks after completion.

The three possible accreditation outcomes are: a three-year accreditation, a one-year accreditation, or a nonaccreditation.

The cost of a survey as of this publication includes a $350 nonrefundable application fee and $726 per day, per surveyor. A typical survey lasts two days and involves two surveyors. The average cost of a survey usually totals $2,900.

STANDARDS TO BE MET

All organizations that apply for CARF accreditation will have the same accreditation conditions, principles, and criteria applied to them. The focus is on the integration of the person served in all areas of the organization and as a decision maker in all aspects of the program. The criteria involve input from people served, ensuring total accessibility (not only architectural but employment and attitudinal accessibility as well as addressing outcomes, safety, and legal requirements).

Facilities or programs also need to meet all of CARF's organizational standards. These address the mission, governance, management, planning, safety, accessibility, information management system, fiscal, personnel, and planning issues of an organization.

The next set of standards applies to the systems that an organization has to have to measure an individual's ability to enter the program; be oriented and assessed; have a plan developed, reviewed, and managed; be referred for additional services; and be discharged and followed after discharge.

Organizations then choose the programs to be accredited based on the specific accreditation section standards. These standards comprise the Medical Rehabilitation Programs: Comprehensive Inpatient Categories One through Three standards (see Exhibit 11.1).

MEDICAL REHABILITATION PROGRAMS

The definition of these programs focuses on their coordinated, interdisciplinary, and integrated services, including evaluation, treatment, education, and training of persons served and their families. The programs are designed to promote outcomes that will minimize and/or prevent impairment, reduce disability, and lessen handicap. The persons served in these programs typically have functional limitations that are caused by disease, trauma, or congenital or developmental impairments. Programs seeking accreditation need to measure their program against the definition to ensure a match between themselves and the standards.

Once it is established that the program description matches what the organization is doing, the process is to compare what is done on a daily basis with the standards. What this entails for some organizations is a totally new approach to how they do business. For others it is a fine-tuning of existing systems and possibly developing some new policies and procedures. Depending on the organization's category, the standards become a guidebook for establishing ongoing quality practices.

CARF standards focus on the patient served, so an organization will need to demonstrate and document its role in a continuum of care that is responsive to those needs. A *continuum of care* as defined by CARF is a system of services providing for the ongoing and/or intermittent rehabilitation needs of persons with significant functional limitations resulting from disease, trauma, aging, and/or congenital and/or developmental conditions. Such a system of services may be achieved by accessing a single provider, multiple providers, and/or a

EXHIBIT 11.1
CARF CATEGORIES

COMMISSION ON ACCREDITATION OF REHABILITATION FACILITIES
COMPREHENSIVE INPATIENT: CATEGORIES ONE THROUGH THREE

CATEGORY	CATEGORY ONE ACUTE	CATEGORY TWO SUBACUTE CARE	CATEGORY THREE SUBACUTE CARE
LOCATION	Hospital	Hospital* Hospital-Based SNF* SNF	Hospital-Based SNF* SNF
MEDICAL NEEDS	Risk for medical instability is high	Risk for medical instability is variable	Risk for medical instability is low
REHAB PHYSICIAN NEEDS	Across all categories regular, direct, individual contact determined by medical & rehabilitation needs		
REHAB NURSING NEEDS	Multiple and/or complex Rehab nursing needs, potential for high medical acuity skilled nursing	Multiple and/or complex Rehab nursing needs, potential for high medical acuity skilled nursing	Routine Rehab nursing needs, low potential for high medical acuity skilled nursing
PT, OT, SPEECH, SS, PSYCH, RT	Should receive, determined by need, minimum three hours	Should receive, determined by need, minimum one to three hours	Should receive, determined by need, minimum one to three hours
AVAILABILITY OF CORE TEAM	5 days a week min.	5 days a week min.	5 days a week min.
OUTCOMES EXPECTED	Progress to another level or community with support as needed	Progress to another level or community with support as needed	Community with support as needed
ED/TRAINING OF PT/FAMILY	Ongoing	Ongoing	Ongoing
EXAMPLES OF PATIENTS BUT NOT LIMITED TO THOSE LISTED	CVA, SCI, TBI, neurological Dxs, multiple trauma, etc.	CVA, multiple trauma, TBI, neuromuscular diseases, etc.	Uncomplicated total hip replacements, amputations, uncomplicated multiple trauma, etc.

*Hospital = unit of larger entity and/or freestanding hospital.   Reprinted with the permission of the Commission on Accreditation of Rehabilitation Facilities.

network of providers. The intensity and diversity of services may vary depending on the functional and psychosocial needs of the persons served.

In an age of healthcare reform and the formation of alliances and networks, CARF has found that these standards push rehabilitation providers to closely examine the needs of the population served and to place patients appropriately within the continuum of care.

CARF standards address issues of ethics by requiring that programs be able to demonstrate their ethical position in the areas of clinical practice, external information about the program, and how they communicate these practices to persons served and their families. Survey teams look for evidence of an underlying structure of ethical practice throughout the survey process. Examples include whether staff are aware of their practice acts and follow them, whether the marketing and collateral material clearly define what the programs are, whether information accurately reflects the services available in the program, and whether patients are educated about their rights.

Medical rehabilitation programs need to clearly state their mission and purpose and need to be able to identify who has responsibility and authority to maintain the operation of the program. This could be more than one individual. The organization should be able to identify for the survey team the title and the responsibility of the responsible individual(s).

CARF has focused for more than 20 years on having programs be responsible for developing outcome systems. The standards address specific outcomes that should be included in the program evaluation system. They focus on efficiency, effectiveness, and consumer satisfaction.

CARF has created standards regarding admission procedures and initial evaluation areas. There is also a standard that encourages clients to tour the program previous to admission when feasible.

Standards address the rehabilitation component as well as the medical components of these programs. There must be the appropriate medical services to meet the pathophysiological process of persons served. The program must also have procedures to obtain emergency medical services.

It is important to have measurable criteria related to timing of initiation and termination of specific rehabilitation services. The survey team will review these criteria and through discussion will ascertain the facility's conformance with the standard.

Standards discuss the environment in which services are offered and the designation of staff to effectively deliver rehabilitation services. For the first time, CARF standards also require medical rehabilitation programs to establish and document a system for determining the types and number of staff members needed by each discipline. There are also standards that address the need for orientation and ongoing training and development of staff, enhancing their interpersonal interactions with the persons served.

Families are a major component of any subacute care program, and standards are in place that include organized educational programs for family members. There should be available information on advocacy, support groups, and assistance with accommodations if necessary. Standards in this area also address the need for programs to have protocols in place to ensure the safety of the patients served. These protocols need to reflect patients' behavioral and cognitive needs.

Follow-up after discharge is a key segment in the program evaluation system. These standards address follow-up in detail. The plan needs to be written and sent to a designated physician and/or service program. Appropriate recommendations that are specific to the needs of the person served must be stated in the report.

The medical rehabilitation program standards also address the issue of pediatric/adolescent programs. There is a two-tiered approach to pediatric/adolescent standards. The field has clearly indicated that there will always be programs that see the occasional child/adolescent but will never consider themselves to be designated pediatric programs. There is a set of standards with which these programs must conform. This is because pediatric populations are not "small adults." They have specific needs that should be addressed in a rehabilitation program. Areas addressed in this first tier focus on the family as pivotal to all areas of the program, the addition of developmental and educational specialists to the core team, the staffing of a pediatric program with qualified staff, and the educational component if these services are needed.

The second tier is for those programs that are designated pediatric programs. In addition to meeting the first tier standards, pediatric programs are also required to have a designated medical director with related experience and training, additional networking capabilities for less common disorders, and a respiratory therapist if children with respiratory problems are being served. The standards will be met by all levels of comprehensive inpatient programs irrespective of the category they have chosen for accreditation.

COMPREHENSIVE INPATIENT: CATEGORIES ONE, TWO AND THREE

According to CARF, *comprehensive inpatient programs* are 24-hour programs that have coordinated and integrated medical and rehabilitation services.

Programs may seek accreditation in a category depending on their licenses held and services offered. A provider may be accredited in more than one category. Categories two and three are for subacute care rehabilitation, whereas category one is reserved for acute rehabilitation programs within facilities licensed as hospitals.

Category two accreditation is available for subacute care rehabilitation programs located in facilities that are licensed as hospitals, hospital-based skilled nursing facilities, or freestanding skilled nursing facilities.

The persons served in these programs have outcomes that focus on returning home or progressing to another level of rehabilitation care such as home health. Persons served in category two are typically individuals with diagnoses that could include but are not limited to: cerebral vascular accidents, neuromuscular diseases, brain injury, orthopedic conditions, and multiple trauma. The program descriptions in the *CARF Standards Manual* outline the necessary components of this category.

The category three designation is for subacute care rehabilitation programs located in facilities licensed as hospital-based skilled nursing facilities or free-standing skilled nursing facilities. The persons served in this category have experienced outcomes of returning to the community with or without support (board and care). Examples of diagnoses could include but are not limited to: orthopedic conditions, amputations, and multiple trauma without complications.

After an organization designates in which category it will seek accreditation, it will be noticed that the differences in standards deal with issues of intensity. The only prescriptive standards that remain address the minimum of therapy hours required from the core treatment team. For category one, therapy must be provided a minimum of five days per week, for a minimum of three hours per day. This is in addition to the education component of an individual's program. For categories two and three, the requirement is that therapy must be provided five days per week, with a minimum of one to three hours of therapy required from the core team, in addition to the educational component of an individual's program.

The standards that cross over all categories identify specific program evaluation measures that include the percentage of unplanned transfers to acute medical facilities and the percentage of patients discharged to long-term care.

The core team is identified and CARF clearly states that the list should not be limited to those staff stated in the standards. As with all CARF standards, members of the core team depend on the needs of the persons served. It should be noted that CARF standards never state that core team members need to be full- or part-time or contracted staff. The survey team will look for the interdisciplinary team's involvement in all areas of decision making and for their commitment to the program. The standards state that core team members should be the primary providers of therapy services and that there should be practice acts for therapies. The team needs to meet in a weekly conference unless it is documented that the needs of the persons served will be met if the conference is every other week.

Many standards address physician issues. For the first time, CARF standards now require that comprehensive inpatient programs have a medical director. The standards outline the requirements to be met for this position. Organizations need to implement a formal credentialing process to determine physician privileges. The standards also outline the rehabilitation physician's role relative to the team. These standards note the amount of physician contact required. The standard for all categories is that the physician will decide on the medical and rehabilitation needs of the person served and the number of direct individual contacts the physician will have with the person served. These contacts should be appropriate to justify the need for continued comprehensive rehabilitation in the category. Physician standards also address concurrent care, the need for physician medical management, and the consultative medical services that must be available (cardiology, family practice, rheumatology, and so on).

The comprehensive inpatient standards for all three categories, also for the first time, address the issue of a director of nursing service in these programs. This individual should not be confused with an organization's director of nursing, who has overall responsibility for all nursing within the organization. The director of nursing service could be the director of nursing but most likely will be the individual responsible for the nursing services in the comprehensive inpatient program, categories one through three. This person promotes competency and staffing levels to meet the rehabilitation and complete needs of the patients. This person develops and implements nursing care plans, ensuring that services are coordinated and providing orientation and training for rehabilitation nursing. Rehabilitation nursing should be available 24 hours per day.

The nursing service in each category must develop an acuity system that will translate into hours of care per person served. The survey team will be assessing whether or not the appropriate intensity of services have been delivered. This should be within 10% of the assessed need that has been established.

CARF emphasizes in the standards for nursing that policies and procedures should demonstrate coordination with other services. Many of these programs may not be based in a hospital, so there are policies and procedures that must be in place to access both emergency medical services and ancillary services available to meet the needs of the populations served.

The programs are to be in designated areas and in an environment that supports the rehabilitation approach to care. There is an understanding that rehabilitation may not be the only program offered. The structure and schedule of the rehabilitation portion should achieve maximum interaction with persons served in rehabilitation. It is important to stress again that *all* categories include these standards.

JCAHO: THE PROCESS

To initiate the process for subacute care accreditation, contact the Joint Commission on Accreditation of Healthcare Organizations to request information and an application. (1) The facility completes the application and returns it with the required deposit. (2) The charge for a subacute care survey averages approximately $2,700 in addition to the charges for surveying the facility. (3) The application asks the facility to designate preferred dates. (4) A JCAHO representative will call the facility to schedule an appointment. It will be at least six weeks before a date is set. (5) Concurrently, JCAHO will send a manual and presurvey material (see Exhibit 11.2), including forms to complete and return. JCAHO will also send a request for information specific to the program (see Figure 11.3), such as the facility's quality assurance plan. Some of this information is sent to the individuals assigned to the subacute care survey team prior to the actual survey date. (6) One-half of the survey fees will be due a month before the survey, with the remainder due within 30 days after the completion of the survey. (7) A scheduler from JCAHO will contact the facility and provide the date of arrival and names of the surveyors. (8) A JCAHO surveyor will call prior to the survey for the purposes of introduction and confirmation of the survey dates. (9) When the surveyors arrive, they will establish an agenda with key department heads. (10) The survey combines interviews, observations, and chart and record reviews. (11) There will be an exit interview upon completion of the survey. (12) The surveyors will complete a report form and send it to JCAHO. (13) The survey is then forwarded to a JCAHO analyst unless a potential condition is indicated, in which case it is sent to the JCAHO unit for special handling. (14) Accreditation is based on a scoring process. A book available from JCAHO entitled *Making Accreditation Decisions* provides more information. (15) The analyst's report is sent to a clinical scan team to determine whether there are any clinical issues to be addressed. (16) If the survey meets the JCAHO standards, the facility will receive a letter regarding its accreditation status. (17) The results of a survey include the following: accreditation with commendation (only about 11% of the total facilities surveyed receive this status), accreditation, accreditation with recommendation (this type requires some type of follow-up procedure), conditional accreditation (this type requires a plan of correction with action within a six-month period), and nonaccreditation.

All accreditation levels are valid for three years. If the subacute care program receives a nonaccreditation standing, it may apply again for a new survey the next day unless the nonaccreditation was because of falsification of records, signatures, or services offered. In this case, reapplication is prohibited for one year.

EXHIBIT 11.2
JCAHO PRESURVEY FORM

Presurvey Subacute Care Database

Facility Name: _____

Address: _____

_____

City: _____ State: _____ Zip _____

1. Ownership:
   Corporate Name _____
   Corporate Address (if different from above) _____
   _____
   _____

2. Facility type (please check one):
   For Profit   ____
   Non-Profit   ____
   Government   ____
   Other (please specify) _____
   _____

3. Our subacute program is operated (please check):
   Within a nursing home   ____
   Within an acute care hospital   ____
   As freestanding   ____
   Within a rehab. hospital   ____
   Other (please specify)_____
   _____

4ai.   Our subacute program operates in a designated unit which has _____ beds.

aii.   Our subacute program does not operate using a designated unit but has
        the capacity for _____ patients.

b.   The information on the matrix covers ____ year(s) or ____ months.

Please complete the attached matrix and return it to:

   Joint Commission on Accreditation of Healthcare Organizations
   Attn: LTCAS
   One Renaissance Blvd.
   Oakbrook Terrace, IL 60181

The survey of subacute programs will be guided by the information contained on this data base
and matrix. Please be as complete as possible in completing the matrix, using patient information from the last year or for as long as the program has been in operation, if less than one
year. We hope, through your participation, to collect valid data on subacute care. Instructions
for completing the matrix are attached. If you have any questions, please call 708-916-5721.

EXHIBIT 11.3
JCAHO—A MATRIX FOR PATIENT INFORMATION

	Patient Categories	Cardiac	Post Surg.	Pulmonary	Renal	Infect. Disease	TBI	Complex Med.	Ortho Rehab.	Other	Other
AT	Total Number of Admits										
AP	Admissions (Total %)										
A1	From Hospital										
A2	From Lower Level Care										
A3	From Emergency Room										
A4	From Rehabilitation										
A5	From Home										
A6	Other										
B1	Length of Stay (Avg.)										
B2	min/max										
C	Interruption in Tx Plan										
C1	Develop Infection										
C2	Sent to Emergency Room										
C3	Expired while in Program										
C4	Other Interruptions										
D1	D/C Hospital										
D2	D/C Home										
D3	D/C Lower Level Care										
D4	D/C Rehab and LTC service										
D5	D/C Other										
E	Payor Source										
E1	Medicare										
E2	Medicaid										
E3	Managed Care (HMO, PPO)										
E4	Self										
E5	Insurance										
E6	Workmen's Compensation										
E7	Other										
F1	Gender Male										
F2	Female										
G1	Age < 12										
G2	13–21										
G3	22–64										
G4	65–84										
G5	85 +										

(Reprinted with the permission of the Joint Commission on Accreditation of Healthcare Organizations.)

EXHIBIT 11.3
(CONTINUED)

Instructions for Completing Matrix

Column Headings:
>   Please enter additional headings that are necessary to describe your functional areas under *"Other"* category in space marked by **** in the example below. List information for *Traumatic Brain Injuries under TBI.*

Patient Categories	Cardiac Rehab	Post Surg.	Pulmonary	Renal	Infect. Disease	TBI	Ortho. Rehab.	Complex Med.	*Other*
									****

A) Admissions:
>   Please indicate the total number of patients in your subacute care program who fall into each of the patient categories. If you provide care for a category of patient(s) not listed, please write in this category in one of the empty columns at the upper righthand corner of the matrix.

>   Example:
>   The ABC Subacute Care Program has been operation for 6 months and has had 100 admissions.
>   Their admissions fall into these groups:
>> 18 Cardiac Rehab Patients;
>> 22 Surgical Recovery Patients;
>> 15 Ventilator Patients
>> 5 HIV Patients;
>> 30 Orthopedic Rehab Patients "includes Stroke patients"; and
>> 10 Stage Four Pressure Sore Patiengs – Complex Medical.

>   ABC Subacute Care Program will report their admissions like this:

	Patient Categories	Cardiac Rehab	Post Surg.	Pulmonary	Renal	Infect. Disease	TBI	Ortho. Rehab.	Complex Med.
AT	Total Number of Admissions	18	22	15	0	5	0	30	10
AP	Admissions (Total %)	18%	22%	15%		5%		30%	10%

The total of the percentages on the admissions row should equal 100%.

Please indicate the number of patients within each category who were admitted from
>> A1 a hospital;
>> A2 a lower level of care (e.g., nursing facility, assisted living, etc.);
>> A3 an emergency room <u>directly</u> to your subacute care program;
>> A4 a rehab hospital or a rehab unit within an acute care or long term hospital;
>> A5 home; and
>> A6 a setting not listed above (please indicate setting, e.g., ambulatory surgery, etc.)..

(Reprinted with the permission of the Joint Commission on Accreditation of Healthcare Organizations.)

EXHIBIT 11.3
(CONTINUED)

Example:

The ABC Subacute Care Program admissions from each category were:

Cardiac Rehab:	All 18 from the hospitsl
Post Surgery:	11 from the hospital, 11 from ambulatory surgery centers
Pulmonary:	15 ventilator admissions from the hospital
Infect. Dis.:	2 from the hospital, 1 directly from an emergency room, 2 directly from home
Ortho Rehab:	20 orthopedic rehab from the hospital, 10 from a rehab unit/hospital
Comp. Med.:	2 from the hospital, 5 from a nursing facility, 1 from a rehab unit/hospital, and 2 directly from home

	Patient Categories	Cardiac Rehab	Post Surg.	Pulmonary	Renal	Infect. Disease	TBI	Ortho. Rehab.	Complex Med.
AT	Total Number of Admissions	18	22	15	0	5	0	30	1
AP	Admissions (Total %)	18%	22%	15%		5%		30%	10
A1	From Hospital	18	11	15		2		20	2
A2	From Lower Level Care								5
A3	From Emergency					1			
A4	From Rehab.							10	1
A5	From Home					2			2
A6	Other		11 (Amb. Surg.)						

B) Length of stay (LOS):

In row B, please indicate the average LOS for each category of patients admitted to your subacute care program within the time frame being reported. Please report LOS in days. This data should pertain to your current census population.

In row B1, please indicate the minimum stay and the maximum stay for each patient category. Please report stays in days.

Example:

The ABC Subacute Care Program reports its length of stay by patient category in Row B and its minimum/maximum stay by petient category in Row B1. The average length of stay for the 10 Complex Medical Patients with pressure sores was 80 days. The shortest stay was 40 days whereas one patient stayed for 140 days.

	Patient Categories	Cardiac	Complex Med.	Post Surg.	Pulmonary	Renal	Ortho. Rehab.	Infect. Disease
B	LOS (avg)	10	80	4	90	0	12	20
B1	min/max	6/20	40/140	2/7	30/10		10/20	5/60

EXHIBIT 11.3
(CONTINUED)

C) Interruption and/or delays in the treatment plan:

Patients in subacute care programs typically follow a defined course of treatment called a critcal pathway. Some patients experience events which interrupt and/or delay the treatment plan. By category, please indicate the number of your admissions who had an interruption and/or delay in their treatment plan because

C1 they developed an infection;
C2 they were sent to the emergency room;
C3 they expired while in the subacute program; and
C4 other interruptions.

In some patient categories, C3 may not be considered an interruption, but an outcome (e.g., hospice); however, please indicate patient deaths at C3. Some patients may need to be counted twice (e.g., developed infection, ultimately sent to the emergency room).

Example:

Patients in the ABC Subacute Care Program experienced the following:

Cardiac Rehab:	3 were sent to the emergency room
Post Surgical:	2 patients expired while in the program
Pulmonary:	5 patients developed infections and 10 patients were sent to the emergency room to assess and/or manage an acute condition change while in the subacute care program
Inf. Disease:	5 HIV patients expired while in the subacute care program
Ortho Rehab:	1 patient expired while in the subacute care program
Complex Med:	2 patients with pressure sores developed infections while in the program

	Patient Categories	Cardiac Rehab	Post Surg.	Pulmonary	Renal	Infect. Disease	TBI	Ortho. Rehab.	Complex Med.
C.	Interruption in TX Plan								
C1	Develop Infection			5					2
C2	Sent to Emergency Room	3		10					
C3	Expired while in Program		2			5		1	
C4	Other interruption								

D) Discharge (D/C) Disposition:

Please indicate by patient category the number of patients discharged to:

D1 an acute care hospital;
D2 home;
D3 a lower level of care (e.g., nursing facility assisted living, personal care home, etc.);
D4 a rehab or long term care hospital; and
D5 other "include deaths."

(Reprinted with the permission of the Joint Commission on Accreditation of Healthcare Organizations.)

EXHIBIT 11.3
(CONTINUED)

Example:

The ABC Subacute Care Program reports its discharges by patient category in the following way:

	Patient Categories	Cardiac Rehab	Post Surg.	Pulmonary	Renal	Infect. Disease	TBI	Ortho. Rehab.	Complex Med.
D1	D/C Hospital	3		10	0				
D2	D/C Home	15	20					19	5
D3	D/C Lower Level Care							10	3
D4	D/C Rehab and LTC Service			5					2
D5	D/C Other		2			5		1	

Include patients who expire (while in the subacute care program) under "D/C Other." Patients whose treatment was interrupted by a visit to the emergency room and were subsequently admitted to the hospital are included in "D/C Hospital."

E) Payor Source:

Please indicate by patient category, the payor sources for your subacute care program.

E1 Medicare

E2 Medicaid

E3 Managed care contract (e.g., HMO, PPO);

E4 Self;

E5 (Insurance – not through a managed care arrangement; and

E6 Other.

Example:

ABC Program reports its payor sources in the following way:

	Patient Categories	Cardiac Rehab	Post Surg.	Pulmonary	Renal	Infect. Disease	TBI	Ortho. Rehab.	Complex Med.
E	Payor Source								
E1	Medicare	3						15	5
E2	Medicaid			15		3			
E3	Managed Care (HMO, PPO)	12	16			1		10	5
E4	Self	3							
E5	Insurance		6			1		5	
E6	Workmen's Comp.								
E7	Other								

(Reprinted with the permission of the Joint Commission on Accreditation of Healthcare Organizations.)

EXHIBIT 11.3
(CONTINUED)

F) Gender

Please indicate by patient category, the gender of your subacute care patients.

Example:

The ABC Subacute Care Program reports the gender of the subacute care patients in the following way:

	Patient Categories	Cardiac Rehab	Post Surg.	Pulmonary	Renal	Infect. Disease	TBI	Ortho. Rehab.	Complex Med.
F1	Gender Male	12	11	10		4		6	4
F2	Female	6	11	5		1		24	6

G) Age

Please indicate by patient category, the number of patients in each age group.

	Patient Categories	Cardiac Rehab	Post Surg.	Pulmonary	Renal	Infect. Disease	TBI	Ortho. Rehab.	Complex Med.
G1	Age < 12		2	3				5	2
G2	13–21		5	2		2		10	1
G3	22–64	6	9	4				10	
G4	65–84	12	5	3		3		5	4
G5	85 +	1	2	3					3

All columns should equal the total admissions to that patient category.

STANDARDS TO BE MET

The Joint Commission on Accreditation of Healthcare Organizations has created a Presurvey Subacute Care Database (refer again to Figure 11.2) to gather data and assess information. JCAHO has also developed a definition and standards for accreditation of subacute care programs. The JCAHO definition is:

> Subacute care is comprehensive inpatient care designed for someone who has had an acute illness, injury, or exacerbation of a disease process. It is goal oriented treatment rendered immediately after, or instead of, acute hospitalization to treat one or more specific active complex medical conditions or to administer one or more technically complex treatments, in the context of a person's underlying long term conditions and overall situation.

> Generally, the individual's condition is such that the care does not depend heavily on high-technology monitoring or complex diagnostic procedures. It requires the coordinated services of an interdisciplinary team including physicians, nurses, and other relevant professional disciplines, who are trained and knowledgeable to assess and manage these specific conditions and perform the necessary procedures. Subacute care is given as part of a specifically defined program, regardless of the site.

> Subacute care is generally more intensive than traditional nursing facility care and less than acute care. It requires frequent (daily to weekly) recurrent patient assessment and review of the clinical course and treatment plan for a limited (several days to several months) time period, until a condition is stabilized or a predetermined treatment course is completed.

The JCAHO standards include requirements that are aimed at enhancing the quality of care offered to subacute care patients. The survey of subacute care programs focuses on chart review with an interest in outcome information as a result of treatment.

The standard of patient rights is addressed. This is accomplished through observation, review of medical records, and discussions with patients. The emphasis is to ascertain the level of the patient's knowledge about his or her condition. Surveyors look for evidence of communication with patients regarding goals and objectives of their care. Surveyors also review documentation of patients' understanding of condition, prognosis, treatment choices, questions, and concerns. For patients who have been discharged, surveyors compare advance directives and request explanations for any disparities.

Leadership is another standard that will be reviewed. Performance is judged primarily from the response and reactions of staff and the patients. The surveyor may use a questionnaire related to the continuous quality improvement process and also employ observation and discussion.

The standard of admission is a part of the survey. The surveyor reviews the preadmission screening and admission decisions in relationship to policies and procedures for admission criteria. Surveyors ask for explanations regarding determination of prognosis and potential recovery on acceptance of a patient and then again at admission. Surveyors may contact referral sources for opinions about the subacute care program's admission process, including any problems related to the facility, the program, or its staff. A policy for refusal to accept a patient and/or a policy regarding a change in patient status upon admission may also be requested.

Surveyors assess case management and case management records. This assessment may include the establishment and management of an interdisciplinary approach, the discharge process, and the coordination of services.

Human resources are another standard that will be reviewed. Staff members may be asked how their skills are assessed and how feedback is given at the facility. The patients and families may be questioned and personnel records may be reviewed.

Standards of patient care will be evaluated. Policies and procedures and specific areas of subacute care will be reviewed. For instance, if the subacute care program admits patients requiring pulmonary rehabilitation, there may be an evaluation of respiratory services. Infection control is an important standard that is addressed. This assessment includes a review of records and policies and procedures. It may also involve discussions with staff, patients, and families.

Patient assessment and evaluation will be reviewed. A policy indicating a time frame for an initial assessment based on the needs of the patient may be requested, as well as a policy for reviewing assessments based on diagnoses. The staff performing the assessments must be qualified and the standards require that their qualifications be documented.

Quality assessment and improvement will be a standard that is included in the survey process. A policy and procedure manual specific to quality may be requested for review. Review of documentation, charts, and records, and discussions with patients' families and staff may be conducted.

Plant operation is also a standard, in addition to the standard of information management. Most information management indicators will be evaluated through actual current and closed cases. For example, discharge summaries should make it possible to understand how the patient reached specific levels at the time of discharge.

The JCAHO standards cover the full scope of subacute care. Policies and/or procedures for the facility may also be appropriate for the subacute care program.

SUMMARY

There is no mystery to becoming accredited. Neither CARF nor JCAHO has any investment in keeping its methodology a secret. In fact, these organizations are more than willing to extend a hand and assist a subacute care program to successfully complete the process.

Both organizations provide meetings, conferences, seminars, manuals, and various materials regarding accreditation. No one will claim that becoming accredited is easy. The facility has to be realistic and plan for a lengthy preparation to ensure that it has all of the necessary mechanisms in place. If the facility or program has never been through the survey process, the administrator may want to seek assistance from consultants who specialize in preparing facilities for accreditation. I would also suggest conducting a mock survey long before the actual one occurs. This mock survey will enable the facility or program to identify its strengths as well as potential problem areas and allow time for correction. It may also help to allay some of the anxiety that staff members may experience due to unknown expectations.

Subacute care is an emerging level of healthcare that is still evolving. With no specific licensure process, subacute care programs depend on the license of the facilities where they are located and the corporation, administrative and clinical staff that develop and implement the program.

Accreditation is the first step on the road to licensure. The benefits of accreditation will serve patients, professionals, providers, and payors.

Standardization of criteria and care delivery with an accepted monitoring system is in the best interests of all clients. The involvement of consumers and professionals in the process will ensure active participation in outcome focused care. A minimum standardized set of skills and services will assist others in determining expectations, definition, and implementation of subacute care.

These efforts toward clarity will be valuable to case managers and payors in selecting appropriate placement and providing reimbursement. The future will undoubtedly bring more regulation; however, we are in the enviable position of being pioneers in the industry and shaping its outcome as it grows and changes.

CHAPTER 12

Integrated
Delivery System Study

*"We should foster the principles of leadership
combined with the spirit of bold curiosity!"*

— Dr. Strangelove

SYSTEM INTEGRATION AND IMPLICATIONS
FOR SUBACUTE CARE

This chapter is a case study about an evolving approach taken by a large healthcare network to design a relevant patient continuum of care management system and supporting organizational structure, while simultaneously transitioning to an integrated service delivery network in an intensely competitive and highly managed care market. The definition and elements of an integrated service delivery model, how subacute care interfaces with the various levels of care, and benefits of an integrated service delivery model will be incorporated.

This case study includes: (1) a description of the environmental planning assumptions and climate that are driving the development of a system integration model; (2) the impact of managed care on shaping and designing a service delivery model; (3) resulting care management systems; (4) the necessary organizational support and an implementation plan for organization change; and (5) lessons learned and future system refinements.

ENVIRONMENTAL PLANNING ASSUMPTIONS

Although there are many different healthcare strategies and models, one constant fact remains: The pressure to reduce healthcare costs is increasing.

A summary of the environmental planning assumptions that are driving the development of a system integrated model include

- Healthcare delivery systems need to cover large populations (regional) and provide all care needs (continuum of care).

- Managed care designed to reduce healthcare costs will continue to expand, creating new reimbursement (capitation) and cost reduction strategies, while reducing hospital admissions and lengths of stay.

- Producing quality outcomes at a competitive cost will determine the facility's ability to secure patients.

- Employees and physicians are active participants in the process of planning delivery and improvement of products and services.

- All services are customer driven and document continuous quality improvement.

- Greatest area of growth will be in subacute care, outpatient niche programs, and at home services.

- Both payors and healthcare systems will continue to consolidate.

Clearly this environment calls for patient care delivery and organizational change. Exhibit 12.1 illustrates relationships among these components of change.

MANAGED CARE IMPACT

Clearly, the pressures creating a drive toward a more managed care environment will continue. Managed care will fundamentally shift the model of healthcare service delivery from one of managing sickness to managing health by modifying incentives for physicians and hospitals. A reduction in the length of stay has been consistently demonstrated in locations with managed care penetration.

A service delivery network can expect changes prompted by managed care. Some of these changes will include an evaluation of patients for appropriateness and cost effectiveness through a continuum of services. Patients will be shifted to lower levels of care and new treatment approaches will be implemented. Traditional rehabilitation and skilled nursing facilities will experience significant changes.

Another change will be the development of aggressive clinical pathways aimed at a greater decrease in the length of stay in acute hospitals. For example, a specific type of total hip protocol may show a reduced hospital length of stay for patients from 4.4 to 3.7 days, with 76% of those patients returning to their homes. A comprehensive, integrated model with aggressive pathways and multiple levels of care will change forever healthcare service delivery patterns, lengths of stay, and costs.

EXHIBIT 12.1
COMPONENTS OF CHANGE

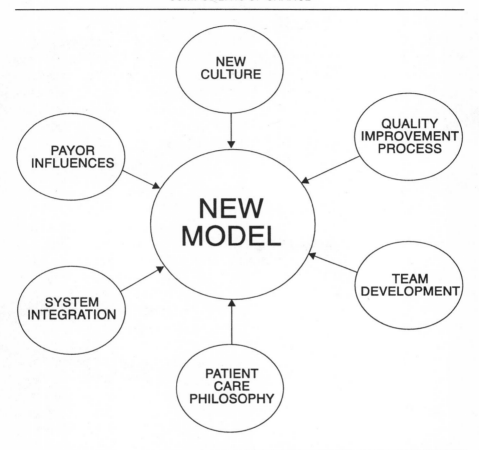

There will continue to be a shifting of patients into lower levels of care, reducing volume in one particular level and increasing in another. This will foster the development of new, integrated, nontraditional approaches.

Additionally, diagnostic mix percentages will change in the area of traditional inpatient acute rehabilitation. A provider with an orthopedic share of their total diagnostic mix needs to anticipate that this level will likely drop as many orthopedic patients shift to alternative levels of care including subacute care and home care therapy. Inpatient rehabilitation programs that have high stroke volume, such as 60% or more, can anticipate losing a percentage of those inpatients to subacute levels of care. For the future, acute inpatient rehabilitation programs will be composed of patients primarily with injuries or illnesses such as traumatic brain injury, spinal cord injury, and recent stroke.

With an increasing managed care environment and payors' desire to channel business to comprehensive healthcare providers, survival for freestanding rehabilitation centers will be limited unless they align, contract, or link with healthcare providers initially in an effort to secure enrollees.

On a more practical level, interaction with the utilization review process will require more energy and effort by internal care managers. This review process now includes managed care physician groups and payors that have local, regional, and national review levels, all with internal or contracted case management and/or utilization review. The provider may experience a situation that involves multiple utilization review and/or case management interactions with all the previously described groups, not always in collaboration. This will increase documentation requirements and necessitate more time for the provider to orchestrate a consistent, collaborative patient care plan among these monitoring forces.

Payors will be working diligently to reduce the cost of service providing little or no increase in payment to healthcare providers. However, the cost of purchased services and goods paid by the providers will continue to increase. If a provider is unable to compete in terms of pricing and with increasing demands for quality improvement, patient referrals will diminish resulting in organizational right sizing and/or closures. Healthcare providers must focus on cost reduction simultaneously with system integration such as alliances, acquisitions, and mergers to remain competitive under a capitation model. The necessity of cost reduction must be understood by all staff throughout the organization.

In this environment, many payors are looking to channel patients to a lower level of care. Simultaneously, the acuity of patients has increased at these lower levels of care, necessitating the hospital-based program to effectively trim costs and the freestanding skilled nursing facility to provide additional services, equipment, and an increased professional nursing and therapy base to deal with these higher acuity-level patients.

CARE MANAGEMENT SYSTEMS

A key landmark of an integrated healthcare model is a comprehensive, coordinated service line with a defined care management system that initiates early intervention, provides low costs, and offers the most appropriate level of care to achieve optimal results. This care management system will need to include a continuum of care with integrated, multiple levels of care, clinical continuum pathways interwoven with defined outcomes, cost and quality indicators.

A system needs to be patient centered and focused to drive the most effective service delivery model. One step in building a continuum model is to establish a lifetime continuum of care framework. A continuum model for persons with physical disability is displayed in Exhibit 12.2.

Services are offered in prevention and wellness, and continue through community and work reintegration. This is critical to the new healthcare paradigm. The structure helps define the components of a service line and conceptually provides an understanding regarding patient flow, service resources, and a sense of partnership with other colleagues practicing in the continuum of care.

Once this inventory of services is defined, it is now possible to establish a four-dimensional integrated continuum care management model. The key elements of the four-dimensional model include: (1) a continuum of services with defined clinical justification, admission, and discharge criteria for each level of care; (2) levels of care and clinical pathways to provide a means of communicating expectations for patient outcomes, to bridge a care continuum, and to promote outcomes within preestablished time frames; (3) functional pathways that define the progression of recovery or adaptation; and (4) quality outcomes that measure a program's effectiveness. The goal of this system is to efficiently manage and coordinate care for patients with disabling conditions across the continuum so that they can reach the highest level of independence and wellness. The focus is a return to the community in the shortest time, and the ability to maintain health/wellness and functional independence.

In a continuum of services, the first component defines all levels of care by diagnosis, admission guidelines, program description, discharge criteria, and so on. This platform of services by level of care creates an opportunity to bring together physicians and clinicians from all levels of care to complete this comprehensive format. For the clinical case manager and others responsible for patient flow, this step is critical for helping the team determine appropriate levels of care. This seamless continuum of care requires a common philosophy and goals be shared by multiple disciplines.

For subacute care providers with both rehabilitation and medical populations, additional levels can be considered given the acuity and diversity of patients. An alternative is to categorize by program, such as ventilator, HIV/AIDS, and wound care. Subacute care programs, staffing, and pricing will be enhanced by carefully designated additional levels of care.

The next dimension is a horizontal, time-driven, diagnostic pathway that overlays across the continuum model. Clinical pathway formats have been successful in managing quality and costs, especially for elective operative procedures. However, applying these same formats for certain other patient groups has been problematic. The typical rehabilitation patient has often suffered from major neurologic injury or illness and presents multiple impairments (bowel, bladder, cognitive, communication, and so on). This is compounded by

EXHIBIT 12.2

AN EXAMPLE OF CONTINUUM OF PROGRAMS AND SERVICES

Prevention and Education	Acute Care	Rehab Care/Acute/ Subacute Care/SNF	Community Reintegration	Work Reentry
• Health screening	• Trauma	• Brain injury	• Outpatient therapies	• Occupational performance
• Professional and community education and seminars	• Orthopedic	• Spinal cord	• Home health	• Vocational evaluation
	• Neurologic	• Stroke	• Sports medicine	• Work evaluation
	• Neurosurgical	• General	• Evaluation Treatment clinics	• Work reentry
	• Other	• Pulmonary	• Support groups	• Training
			• Peer counseling	
		With the following services:	• Maintenance groups	
			• Day treatment	
		PT, OT, Speech, Nursing, Psych, Social Work, Recreation Therapy, Case Management, and Discharge Planning	• Transitional living	
			• Community and home	
			• Disability assessment	
			• Pain rehab	

variable recovery rates and improvements. Impairments are typically chronic, complicated, and costly to treat.

These impairments lead to functional disability that needs to be addressed by multiple levels of care across the continuum model. Therefore, clinical pathways need to be designed on a continuum basis rather than just one particular level of care to fully ensure integrated patient care delivery.

A methodology that evaluates functional abilities (such as self care, mobility, and communication) across the parameters of recovery or adaptation including dependence to independence and measured numerically is the third dimension of this model. For persons with physical disability, a system needs to be adopted as the common platform for documenting functional change.

An added advantage of utilizing a common system is that it then allows various disciplines, for example physical therapy, to standardize mobility procedures and techniques. Treatment guideline results may be used by the care provider to render the treatment for a specific diagnosis and functional impairment. Treatment guidelines provide recommendations and instructions for the care of all diagnoses resulting in major disabilities for each of the areas of functional impairment and level of impairment. The treatment guidelines can be used throughout the system, allowing a smoother transition of patients throughout the various levels of care. They will also encourage appropriate care to be introduced at the earliest time, ensuring common practice standards for treatment throughout the continuum, and also ensuring quality and consistency at each service delivery location.

The fourth dimension to this care management system is quality and outcome measurements, primarily efficiency and effectiveness outcomes, defined across the continuum and per level of care. These data enable team members and program designers to address issues regarding timeliness, quality, and cost-effective practices. Efficiency might include lengths of stay, charges/costs per day, and improvement change per day. Effectiveness would encompass the percentage of clients discharged to home, functional level changes from admission through discharge, and patient satisfaction.

The data allow the members of the continuum as a whole to ask questions and implement new programs and services collaboratively. For example, a treatment team at a subacute care level may ask whether costs can be achieved and quality maintained by reducing the patients' length of stay. This, again, is not done in isolation, but with colleagues functioning in all other levels of care, ensuring that the entire program has the services and methodologies to address the modified pathway.

A resulting database will provide increased accuracy for predicting outcomes and the most appropriate care track to achieve a patient's goals. It is consistent with an integrated network's responsibility for managing the lifetime healthcare needs of enrollees to provide a base for continuous staff education, standardized tools and techniques, language, and measurement of patient outcomes.

Implementing a service line of care, however, can cause concern on the part of stakeholders in a particular level of care. Acute inpatient personnel can become alarmed with the development of subacute care services. Being part of a development team that completes the continuum of care format will enable them to see that each level has a particular role in the system. A profile of the types of patients being served in subacute care, long term care, and acute care helps dispel the notion that subacute care is a competitor.

ORGANIZATIONAL SUPPORT MODEL

Hospital organizations are rapidly transforming through consolidations, mergers, and alliances to position themselves as integrated healthcare networks. A checklist of integrated healthcare system characteristics can be found in Exhibit 12.3.

Without a consistently applied philosophy and plan, integrating multiple disciplines and entities becomes difficult. Two concepts relevant to system integration are: (1) service lines that are responsible for the full continuum of care, specific programs, or centers of excellence, using clinical pathways/algorithms and partnerships with physicians and teams to manage patients; and (2) centralized departments accountable for providing resources to the service lines at all levels of the continuum and to all geographic sites.

For instance, with this model one therapy division would exist so that there are no longer separate therapy departments at multiple locations. The single-service delivery department provides services at multiple locations, obviously influenced by the environmental, cultural, organizational, and physician context at each particular location. A patient matrix organizational service delivery model results. One way of envisioning this model is through the analogy of a hospital or therapy location as an airport with multiple medical groups/prepaid plans (the airlines), with patients (passengers), landing at various airport gates (to access programs and services) that are standardized at the various airport locations.

An interface structure is necessary to assemble a variety of multidisciplinary, multientity service line groups; to provide input to approve and monitor business plan goals; and to integrate, systematize, and consolidate as appropriate. It is important to ensure interdisciplinary, multientity interface through communication and collaboration by involving all of the stakeholders and increasing patient care team empowerment and accountability for improving performance and outcomes.

EXHIBIT 12.3
AN INTEGRATED HEALTHCARE SYSTEM CHECKLIST

INTEGRATED HEALTHCARE SYSTEM CHECKLIST

Organization

✓   Single operating company or network of corporate alliances

✓   Single board of directors

✓   One "bottom line"

✓   Improve healthcare of community

✓   Physician groups/networks are core of system

Patient Care

✓   Service line with lifetime continuum models

✓   Centralized system departments/programs

✓   Interdisciplinary program teams

✓   Strong linkage with managed care physician groups

✓   Emphasis on multispecialist skills and group performance

✓   Standardized procedures and system with practices

✓   Wellness and prevention services

✓   Integrated electronic documentation systems

✓   Outcome oriented patient planning

✓   Customer driven

A steering committee would serve as the board with key physician and system leadership, meeting on a bimonthly or quarterly basis to monitor the progress and activities of the service line. Steering committee membership may consist of the following individuals:

- Medical directors
- Quality assurance
- Financial
- Marketing
- Contracting specialists

- ◆ Operations providers (i.e., nursing, therapy)
- ◆ Medical group representation
- ◆ Other service line and/or level of care representatives (orthopedic/neurologic, subacute care, and home health, for example)

THE CASE FOR CHANGE

Moving large groups of multidisciplinary individuals traditionally housed in discipline departments fragmented by separate entity locations to an integrated purpose requires an orchestrated message. You must stress the changing healthcare delivery paradigm. Explain the differences between managed care and capitation and identify the attributes of future healthcare service delivery systems. It also helps to define continuous quality improvement, and work redesign and reengineering. Be prepared to discuss the system-specific strategies to respond to a local/regional changing healthcare market.

The system message needs to be given repeatedly to employees and physicians to prepare them for the full impact and ongoing change that they will encounter in healthcare. Staff must have a working knowledge and vocabulary of managed care and capitation. They should understand how the money flows, creating new incentives for wellness and reduced hospitalization.

Exhibit 12.4 lists examples of the paradigm shifts.

Staff should understand the payor mix, providers, and healthcare market conditions occurring in the local environment. It is necessary for staff to understand that these conditions are affecting the entire healthcare community, not just one particular department or hospital. Staff members need to understand the concepts of horizontal integration, the need for geographic coverage, and the impact of capitation.

All staff should know that hospitals will compete on costs and win on quality. All stakeholders—physicians, nurses, therapists, patients, and families—need to be a part of design change. Right sizing to a realistic census, not only in terms of hospital workers but also the ratio of primary practice to specialty physicians, will occur. A variety of methodologies for right sizing include work redesign and reengineering. This message needs to be carried by the leaders of the organization, in writing and by face-to-face meetings to communicate the message, receive feedback from staff, and solicit ideas to facilitate change. Be prepared for this message to significantly stir the organization. Anticipate questions from physicians and staff. Examples of possible questions are: Will the need for reductions in hospital beds affect me? What has happened to the census over the past few years? How will quality be maintained and measured? Is the

EXHIBIT 12.4
COMPARISON OF PREVIOUS AND EMERGING HEALTHCARE MODELS

OLD	NEW
PAYOR INFLUENCE	
1. Indemnity/Medicare	1. Managed care
2. Admission/high census	2. Covered lives
3. Fee for service	3. Capitation
TEAM DEVELOPMENT	
1. Traditional departments	1. Self-managed program teams/patient focus
2. Directed teams	2. Participative leadership
3. M.D./team separation	3. M.D./team participation/integration
4. Individual performance	4. Team/group performance
5. Specialist	5. Family practice/multispecialist
PATIENT CARE PHILOSOPHY	
1. Treat illness	1. Prevention/wellness
2. Change patient	2. Change environment and systems
3. Professional directed plan	3. Professional in supportive role to consumer
QUALITY IMPROVEMENT	
1. Mono culture	1. Cultural diversity
2. Hospital internal orientation	2. Customer service
3. Department quality	3. Total quality culture
SYSTEM CONTINUUM INTEGRATION	
1. Separate entities	1. Program integration
2. Hospital centered	2. Community integrated

3. Inpatient emphasis	3. Continuum of care
	  ◆ Develop community supports
	  ◆ Acute care
	  ◆ Subacute care
	  ◆ Skilled nursing facilities
	  ◆ Outpatient/niche programs
	  ◆ Home health

NEW CULTURE	
1. Low visibility	1. High visibility
2. Conservative admissions/three key programs	2. Flexibility, diverse programs
3. Defend the past	3. Look to new learning situations

organization willing to help me through this transition? Who is involved in redesigning the organization? What opportunities exist for input?

Remember to communicate and educate: this openness will prepare the organization for change.

SUMMARY

The system with an experienced program will be able to share staff, program development opportunities, cross train, and so on, as changes occur. Support departments will need to be educated regarding the myriad rules and regulations that accompany a service delivery program.

For example, staff need to realize that in a skilled nursing facility the patient is not called *a patient*, but a *resident*. Staff will need to ensure that they knock on the door and provide other courtesies in recognition that the room is the patient's home.

A critical component in developing a subacute care program is to exercise extreme caution in selecting nursing personnel. A critical care or acute medical surgical nurse, or one who has long-term care experience, does not necessarily indicate a successful match with the subacute level of care. One thing is certain: long-term care nurses must be supplied with a skill development program to meet the increased acuity needs.

Interdisciplinary teamwork is critical to ensure that functional patient goals are achieved in a short period of time. The hallmark of an interdisciplinary team is that all members reinforce and collaborate on patient treatment goals. For example, the nursing staff are not only responsible for nursing goals but also need to reinforce the plan developed by the therapy department.

Physicians should understand that they can continue to follow patients on the subacute care unit and will need to do so more frequently than once per month. The skilled nursing facility level of care brings with it a host of rules and regulations that involve additional justification of prescriptions, especially in the area of behavioral management using psychotropic medications. The traditional long-term care medical director is passé. The medical director will need to be highly involved in patient program development, physician interaction, and the managing and monitoring of quality patient care. Additional specialists will be seen more frequently in subacute care/skilled nursing facilities offering and managing services.

Changing levels of care must be accompanied with an education process to modify the expectations of patients and families. Patients and families need to understand that the number of registered nurses and the number of physician visits may be different. The length of stay will be based on the functional progress of the patient.

It is critical that the provider investigate insurance benefits so that an individual who does not have certain benefits will be able to make appropriate arrangements. Often one hears patients and families indicate "but I have excellent insurance"; however, the plan does not contain an adequate skilled nursing facility and/or custodial benefit.

In the future, acute care facilities will become much more cost conscious and competitive. They will compete directly with subacute care programs and skilled nursing facilities. Within a system integrated approach healthcare facilities will be asked to provide guidance as well as management.

Regulatory influence such as the three-hour therapy rule in acute rehabilitation units under a managed care environment will not be that critical. Patients will be placed in rehabilitation programs that may include acute rehabilitation and subacute care levels dictated by the individualized patient functional treatment goals, not an artificial three-hour rule for therapies.

Subacute care program growth will be a result of patient shifting, replacing medical/surgical beds, and not expanding new beds in a healthcare system. Realtime cost accounting and pathway variance reporting will provide outcomes to empower teams and enhance the care management process.

The healthcare change will continue, creating an exciting opportunity for competent leaders. Building a system integrated model will require vision, a crafted plan, and a great deal of determination. Traditional methods and incentives will continue to change how we provide services. A healthcare organization's ability to respond will determine its ability to survive.

Conclusion

"Yes."
— Dr. David Livingston

Subacute care is an approach to care that offers outcome-focused services. The patient and his or her family become an integral part of the treatment team. The interdisciplinary team is composed of professionals who are experienced in their specialty area.

Care is provided in a more personal atmosphere with consideration given to continuity. In the subacute care program the patient encounters the same team members each day and is encouraged to actively participate in the care planning process. The nurses, case manager, social worker, therapists, physician, and others, are available to meet frequently regarding goals and objectives. It is common for the subacute care patient to know the dietary staff, the clinical staff, housekeeping personnel, and the schedule of activities. The patient knows that a family member or loved one can share a meal or participate in therapies and recreational activities. In some instances, if appropriate, family members may spend the night.

The initial appeal of the subacute care program is that cost-effective and appropriate healthcare can be delivered. The true beauty of a subacute care facility is that it offers a place to recover in a small comfortable setting where the patient is as much a part of the interdisciplinary team as is the physician. The subacute care model has restored some humanity to healthcare and made inpatient confinement significantly more comfortable.

Unlike some settings where attention is placed on the pathology of the illness or injury, the subacute care program treats the full scope of the patient and family and the roles they seek to preserve. Although reimbursement may have stimulated the development of subacute care programs, service is the quality that will enable subacute care programs to grow at a more remarkable rate. Consumers are becoming better educated regarding healthcare options, and they are already showing a preference for subacute care settings.

Today subacute care programs are the fastest growing segment of the post-acute service market; tomorrow they will be commonplace. Several factors are driving this rapid growth. Payors are seeking the least expensive service level that can achieve desired outcomes. Managed care payors benefit immediately by moving patients into lower cost settings that can deliver quality services. The "grayby boom" is a powerful influence demanding an increase in post-acute care services. The over 85-year-old segment of the population is rising at a rapid rate of 4% per year. There are regulations in the form of certificates of need in some states that are restricting the supply of certain types of healthcare facilities. The movement of payors to impose greater controls on healthcare providers is stimulating consolidation. There is an effort to establish networks of integrated delivery systems that can manage patient costs and become more attractive as a coordinated system to managed care referral sources.

Finally, in an age of capitation, there is an opportunity to reduce risk by moving the patient to the least costly but most appropriate setting.

As we move into the next century, subacute care will be a significant influence in redefining American healthcare.

PART IV

◆

APPENDICES

Acronyms and Abbreviations

AAMC Association of American Medical Colleges
AAPA American Academy of Physician Assistants
AAPCC Adjusted Average Per Capita Cost
AARP American Association of Retired Persons
ABMS American Board of Medical Specialists
ACP American College of Physicians
ACRM American Congress of Rehabilitative Medicine
ACS American College of Surgeons
ADAMHA Alcohol, Drug Abuse, and Mental Health Administration
ADL Activities of Daily Living
ADS Alternative Delivery System
AHA American Hospital Association
AHC Academic Health Center
AHCA American Health Care Association
AHCPR Agency for Health Care Policy and Research
AHEB Adjusted Historical Education Basis
AHPB Adjusted Historical Payment Basis
AMA American Medical Association
ANA American Nurses Association
ASCA American Subacute Care Association
ASPE Assistant Secretary for Planning and Evaluation
BDMS Bureau of Data Management and Strategy, HCFA
BLS Bureau of Labor Statistics
BMAD Part B Medicare Annual Data Files

BPO Bureau of Program Operations, HCFA
CARF Commission on Accreditation of Rehabilitation Facilities
CBO Congressional Budget Office
CBS Current Beneficiary Survey
CCM Certified Case Managers
CCMC Committee on the Costs of Medical Care
CF Conversion Factor
CHAMPUS Civilian Health and Medical Program of the Uniformed Services
CHAMPVA Civilian Health and Medical Program of the Veterans Administration
CHER Center for Health Economics Research
CMP Community Medical Plan
CMSA Case Management Society of America
COBRA Consolidated Budget Reconciliation Act
COGME Council on Graduate Medical Education
CON Certificate of Need
CPR Customary, Prevailing, and Reasonable (fees)
CPT Current Procedure Terminology
CR Capitated Rate
CRS Congressional Research Service
CWF Common Working File
DEFRA Deficit Reduction Act
DRG Diagnosis Related Group
DSM Diagnostic and Statistical Manual
ECF Extended Care Facility
EOMB Explanation of Medical Benefits
EPO Exclusive Provider Organizations
EPSDT Early and Periodic Screening, Diagnosis, and Treatment
ESP Economic Stabilization Program
FFS Fee for Service
FHSR Foundation for Health Services Research
FMAP Federal Medical Assistance Percentage
FTE Full-Time Equivalent
GAF Geographic Adjustment Factor
GAO General Accounting Office
GDP Gross Domestic Product

GFMPC Geographic Adjustment Factor for Malpractice Costs in Locality
GHAA Group Health Association of America
GHC Group Health Cooperative
GMENAC Graduate Medical Education National Advisory Committee
GNP Gross National Product
GPCI Geographic Practice Cost Index
HB/SNF Hospital-Based Skilled Nursing Facility
HCFA Health Care Financing Administration
HCPCS Common Procedure Coding System, HFCA
HHS Department of Health and Human Services
HIAA Health Insurance Association of America
HIO Health Insuring Organization
HIP Health Insurance Plan
HIS Health Interview Survey
HMO Health Maintenance Organization
HPDP Health Promotion and Disease Prevention
HSQB Health Standards and Quality Bureau, HCFA
HSUS Health Services Utilization Study
IADL Instrumental Activities of Daily Living
ICD International Classification of Diseases
ICMA Individual Case Management Association
IPA Independent Practice Association
JCAHO Joint Commission on Accreditation of Healthcare Organizations
LCL Lowest Charge Level Limit
LLP Limited License Practitioner
LTC Long-Term Care
MAAC Maximum Allowable Actual Charge
MCO Managed Care Organization
MDE Maximum Dollar Expenditure
MDS Minimum Data Set
MEI Medicare Economic Index
MFS Medicare Fee Schedule
MGMA Medical Group Management Association
MMPS Medicare Mortality Predictor System
MOS Medical Outcomes Studies
MPC Malpractice Costs Relative Value for Service

MSA Metropolitan Statistical Area
MSO Management Services Organization
MUA Medically Underserved Area
MVPS Medicare Volume Performance Studies
NACHM National Advisory Commission on Health Manpower
NCHS National Center for Health Statistics
NCHSR National Center for Health Services Research
NCOA National Council on the Aging
NHDS National Hospital Discharge Survey
NHSC National Health Services Corps
NICHMOD National Industry Council for HMO Development
NIH National Institutes of Health
NLT National Long-Term Care
NMCES National Medical Care Expenditures Survey
NMCUES National Medical Care Utilization and Expenditures Survey
NMES National Medical Expenditures Survey
NNHS National Nursing Home Survey
OACT Office of the Actuary, HFCA
OBRA Omnibus Budget Reconciliation Act
OEO Office of Economic Opportunity
OMB Office of Management and Budget
OPC Office of Practice Costs Relative Value for Services
OSHA Occupational Safety and Health Act
OTA Office of Technology Assessment
PAR Participating Physician and Supplier Program
PCF Patient Compensation Fund
PDR Per Diem Rate
PGP Prepaid Group Practice
PHP Prepaid Health Plan
PHS Public Health Service
PLI Professional Liability Insurance
POS Point of Service
PPO Preferred Provider Organization
PPRC Physician Payment Review Commission
PPS Prospective Payment System
PPSIC Physicians' Practice Cost and Income Survey

PRO Peer Review Organization
ProPAC Prospective Payment Assessment Commission
PTW Physicians' Total Work Relative Value for Service
QA Quality Assurance
QMB Qualified Medicare Beneficiary
RAPs Radiologists, Anesthesiologists, and Pathologists
RB-RVS Resource-Based Relative-Value Scale
RHC Rural Health Center
RVU Relative Value Unit
SEM Standard Error of the Mean
SHMO Social Health Maintenance Organization
SMI Supplementary Medical Insurance Program
SNF Skilled Nursing Facility
SOA Supplement on Aging
SSI Supplemental Security Income
TEFRA Tax Equity and Fiscal Responsibility Act
UCR Usual, Customary, and Reasonable
UR Utilization Review
URVG Uniform Relative Value Guide
VA Veterans Administration
VE Voluntary Effort
VPS Volume Performance Standard

Glossary

Actuary: an insurance professional who mathematically analyzes and prices the risks associated with providing insurance coverage.

Adjusted Community Rating: a rate-setting methodology based on group specific demographics.

Administrative Services Only: a contract with an insurer to perform administrative service only for a self-insured organization.

Affiliated Provider: a provider or organization who is subcontracted by an HMO to provide services to members.

All-Payor System: a rate-setting program whereby all third-party payors pay the same rates, set by the government, for the same medical services.

Alternate Delivery System: alternatives to traditional health care financing and processes for providing medically necessary services in a more cost-efficient manner such as an HMO or PPO.

Ambulatory Care: Health care services or medical procedures performed on an outpatient basis whereby the patient is able to return home without an overnight stay in a medical facility.

Ancillary Services: supplemental services provided to patients other than room and board and nursing. Such services could include x-ray, laboratory work, MRI, cardiac testing, physical therapy, occupational therapy and speech therapy.

Average Length of Stay (ALOS): the average number of days that a patient remains in the hospital for a given time period.

Benefits: the money or services provided under the terms of an insurance policy.

Board Certified: a physician who has passed examinations by a professional association that regulates his specialty.

Capitation: a per-member, monthly payment to a provider which covers contracted services and is paid in advance of the delivery of service.

Case Management: the monitoring and coordinating of treatment for specific diagnosis, particularly high-cost or extensive services.

Claim: a statement of health services and their costs provided by a hospital, physician's office or other provider facility.

Coalitions: association of healthcare sponsors that pool resources to gather information on and negotiate with insurers and other healthcare providers.

Coinsurance: the percentage of the costs of medical care that a patient pays himself.

Community Rating: premiums based on the average cost of providing medical services to all people in a geographic area without adjusting for an individual's medical history or the likelihood of using such services.

Concurrent Review: a review of a procedure to ensure that medically necessary and appropriate care is delivered during a patient's hospital stay.

Co-Payment: a cost-sharing arrangement in which the HMO enrollee pays a specified fee for a specific service.

Continued Stay Review: a utilization review technique that monitors the continued appropriateness of a member's hospital stay.

Coordination of Benefits (COB): a cost control measure to prevent a covered person from receiving duplicate benefits.

Cost-Based Reimbursements: a method of paying providers an amount which is based on the cost to the provider of delivering services.

Deductible: the amount of money a person must pay before the payor begins reimbursement.

Department of Health and Human Services: the cabinet agency of the U.S. government which administers most federal health programs.

Diagnostic Related Group (DRG): an inpatient classification system used by the government in which hospital procedures are rated in terms of cost, after which a flat rate is set per procedure.

Discharge Planning: a review process in which healthcare professionals identify and evaluate the anticipated needs of a patient following discharge from the hospital.

Dual Diagnosis: more than one diagnosis.

Enrollee: member or beneficiary of a health plan.

Enrollment: the number of members in an HMP/PPO or other healthcare organization.

Exclusive Provider Organization (EPO): similar to a preferred provider organization in organization and purpose; however, it limits enrollees to receive all of their covered services from providers that participate in EPO.

Exclusions: Specific conditions listed for exclusion in a contract.

Fee: a charge or price for professional service.

Fee-for-Service: a system of payment for health care where a fee is charged for each service provided.

Fee Reimbursement: the method of payment received for services rendered to patients by an HMO after the care has been provided.

Fee Schedule: a listing of charges or established benefits for specific medical or dental procedures.

Gatekeeper: the primary care physician or manager (case manager) who must authorize all medical services.

Global Budget: term used to place a nationwide limit on overall spending for healthcare services.

Grace Period: a period of time after a premium is due during which coverage must be provided and the premium paid without penalty.

Group HMO: an HMO that contracts with one or more independent group practices to provide services in one or more locations, in which the physicians of the group are paid on a capitated basis.

Health Care Financing Administration (HCFA): the agency within the Department of Health and Human Services which administers federal health financing and related regulatory programs, principally the Medicare and Medicaid programs.

Health Maintenance Organization: an organized healthcare system that is responsible for both the financing and delivery of a broad range of comprehensive healthcare services to an enrolled population for a prepaid, fixed fee.

Health Insurance Purchasing Cooperative: per President Clinton's restructured health system, they are purchasing agents for large groups of employers in a region, which would shop for the highest-quality health plan at the lowest price.

Home Health Care (HHC): medical care provided by certified professionals in the patient's home for patients who no longer need the extensive treatment provided by a hospital or skilled nursing facility.

Hospice Care: an organized program that provides palliative and supportive care for terminally ill patients at home or in a hospice rather than a hospital.

Incurred but Not Reported: to be liable for a loss, claim or expense which occurred within a fixed period and for which an insurance company or medical provider becomes liable whether or not yet reported, adjusted and paid.

Indemnity: protection against loss. Also in reference to private insurance.

Independent Physician Association: contracts with individual physicians who see patients for a set rate, as well as their own patients, in their own private offices.

Individual Practice Association: One of four models which is a mixture of physicians or healthcare professionals from solo and group practices.

Insured Services: special procedures or tests performed on a patient which are considered part of the services to be covered.

Long-Term Care (LTC): the services that would be required over an extended period of time by someone with a chronic illness or disability, usually in a skilled nursing facility.

Managed Care: a system of managing and financing health care delivery to ensure that services provided to plan members are necessary, efficiently provided, and appropriately priced.

Medicaid: state programs of public assistance to persons whose resources are insufficient to pay for medical care.

Medicare: a federally funded nationwide hospital and medical care insurance program for citizens over age 65 and for disabled people of all ages.

Medicare—Part A: covers inpatient costs for Medicare patients.

Medicare—Part B: covers ancillary and outpatient costs for Medicare patients.

Medicare Supplement: an insurance program that specifically covers those costs not covered by Medicare.

Member: same as enrollee or beneficiary. A person eligible to receive benefits from an insurance policy.

Morbidity: an actuarial concept that shows the average incidence of illness occurring in a large group of people.

Multi-Specialty Group: a group of doctors who represent various medical specialties and who work together in a practice.

Negotiated Fee Schedule: a payment system in which providers work for a fixed portion of their usual fee.

Network HMOs: a model which contracts with two or more independent group practices to provide health services.

Open Enrollment: the annual period during which people can choose from health plans being offered.

Out-of-Area: refers to the treatment given an HMO member outside the geographical limits of his own HMO.

Outpatient Surgery: minor ambulatory surgery.

Per Review: a cost and quality-control system in which physicians review or evaluate the work of other physicians using predetermined standards.

Penetration: the percentage of business an insurer is able to capture in the market areas as a whole.

Per Diem Cost: cost per day in a facility providing medical care.

Per Member per Month (pmpm): the fee paid under a capitated contract to the provider for members enrolled in a plan and which is paid on a monthly basis.

Point-of-Service: a plan which offers participants the option of using panel providers or nonpanel providers of the participant's choosing for an additional fee.

Preexisting Medical Condition: a physical and/or mental condition that a patient has before applying for insurance coverage.

Preferred Provider Organization (PPO): a group of providers that contracts on a discounted fee-for-service basis to provide services to subscribers.

Preventive Health Care: care that has as its goal, prevention of disease before it occurs and concentrates on keeping patients well in addition to healing them when they are sick.

Primary Care: the initial treatment received when an individual first seeks medical care.

Primary Care Case Manager: a program which offers primary care providers the opportunity to provide and assume risk for physicians and selected outpatient services and case management of inpatient and other services on a nonrisk basis.

Primary Care Physician: the physician who assumes the responsibility for the comprehensive care of an individual on a continuing basis.

Prospective Payment System: a standardized payment system implemented in 1983 by Medicare to help manage health care reimbursement whereby the incentive for hospitals to deliver unnecessary care is eliminated.

Quality Assurance: a set of activities to assure the quality of services provided.

Reinsurance: the practice of one insurance company purchasing insurance from another company to protect itself against part or all of the losses incurred in the process of honoring the claims of policyholders.

Residential Care Facility (RCF): a facility that provides persons with food and shelter and a limited amount of other services.

Resource-Based Relative Value Scale (RBRVS): a system for determining the level of payments to physicians based upon the amount of work involved in the treatment.

Risk Contract: a contract in which all parties involved are at risk.

Risk Sharing: the process whereby an insurer and contracted provider each accept partial responsibility for the financial risk and rewards involved in cost effectively caring for members.

Safety Net Provider: one that provides services to the medically indigent and special needs segments of a state's population.

Self-Insurance: a program providing health benefits in which a business pays the cost of employees' medical benefits up to a certain amount.

Senior Plan: benefits package to the non-working person over the age of 65, often under Medicare through a contract with the federal government.

Service Area: the geographic area served by an insurer or healthcare provider.

Shared Risk Pool: incentive program for controlling the cost of selected services when actual incurred costs are favorable compared to a predetermined budget; the savings may be shared with the physician, the hospital, and others.

Single Payor: a system whereby one entity pays for all health care.

Skilled Nursing Facility (SNF): a facility that provides skilled nursing services.

Specialty HMO: an HMO organized around a specific medical specialty.

Staff-Based HMO: an HMO the services of which are delivered at one or more locations by doctors who are employed by the HMO.

Stop-Loss: the practice of an insurance company protecting itself or its con-tracted medical groups against part or all losses above the specified dollar amounts.

Supplemental Health Services: optional services that an insurer might pro-vide in addition to the basis care, usually at additional cost.

Targeted Case Management (TCM): project which involves intensive case management of high-cost or high-risk beneficiaries to try to positively impact patient outcomes.

Third-Party Administrator (TPA): an organization or person who provides certain administrative services to group benefits plans, including premium accounting, claims review and payments.

Third-Party Payment: payment for healthcare services by someone other than the individual who received the care or administered it.

Unbundled Services: the provision of a wide range of healthcare services that may be purchased separately.

Utilization Review: a cost-control mechanism used in recent years to evaluate health care on the basis of appropriateness, necessity and quality.

Ventilator Dependent: a person who is dependent on a ventilator to enable them to breathe. It may be continuous or intermittant.

Ventilator Weaning: Activities that permit the patient to no longer require mechanical means to breathe.

Withhold: the portion of the monthly capitation payment to providers withheld until the end of the year to create an incentive for efficient care.

Worker's Compensation: a state-operated insurance program that protects against loss of employee income by providing benefits to employees who are injured on the job.

Wraparound: a supplementary plan designed to pay for benefits not provided under the basic plan.

Index